The Open University

ROYAL SOCIETY OF CHEMISTRY

The
Molecular World

The Third Dimension

edited by

Lesley Smart and Michael Gagan

This publication forms part of an Open University course, S205 *The Molecular World*. Most of the texts which make up this course are shown opposite. Details of this and other Open University courses can be obtained from the Call Centre, PO Box 724, The Open University, Milton Keynes MK7 6ZS, United Kingdom: tel. +44 (0)1908 653231, e-mail ces-gen@open.ac.uk

Alternatively, you may visit the Open University website at http://www.open.ac.uk where you can learn more about the wide range of courses and packs offered at all levels by The Open University.

The Open University, Walton Hall, Milton Keynes, MK7 6AA

First published 2002

Edited, designed and typeset by The Open University.

Published by the Royal Society of Chemistry, Thomas Graham House, Science Park, Milton Road, Cambridge CB4 0WF, UK.

Printed in the United Kingdom by Bath Press Colourbooks, Glasgow.

ISBN 0 85404 660 7

A catalogue record for this book is available from the British Library.

1.1

s205book 3 i1.1

The Molecular World

This series provides a broad foundation in chemistry, introducing its fundamental ideas, principles and techniques, and also demonstrating the central role of chemistry in science and the importance of a molecular approach in biology and the Earth sciences. Each title is attractively presented and illustrated in full colour.

The Molecular World aims to develop an integrated approach, with major themes and concepts in organic, inorganic and physical chemistry, set in the context of chemistry as a whole. The examples given illustrate both the application of chemistry in the natural world and its importance in industry. Case studies, written by acknowledged experts in the field, are used to show how chemistry impinges on topics of social and scientific interest, such as polymers, batteries, catalysis, liquid crystals and forensic science. Interactive multimedia CD-ROMs are included throughout, covering a range of topics such as molecular structures, reaction sequences, spectra and molecular modelling. Electronic questions facilitating revision/consolidation are also used.

The series has been devised as the course material for the Open University Course S205 *The Molecular World*. Details of this and other Open University courses can be obtained from the Course Information and Advice Centre, PO Box 724, The Open University, Milton Keynes MK7 6ZS, UK; Tel +44 (0)1908 653231; e-mail: ces-gen@open.ac.uk. Alternatively, the website at www.open.ac.uk gives more information about the wide range of courses and packs offered at all levels by The Open University.

Further information about this series is available at www.rsc.org/molecularworld.

Orders and enquiries should be sent to:

Sales and Customer Care Department, Royal Society of Chemistry, Thomas Graham House, Science Park, Milton Road, Cambridge, CB4 0WF, UK

Tel: +44 (0)1223 432360; Fax: +44 (0)1223 426017; e-mail: sales@rsc.org

The titles in *The Molecular World* series are:

THE THIRD DIMENSION
edited by Lesley Smart and Michael Gagan

METALS AND CHEMICAL CHANGE
edited by David Johnson

CHEMICAL KINETICS AND MECHANISM
edited by Michael Mortimer and Peter Taylor

MOLECULAR MODELLING AND BONDING
edited by Elaine Moore

ALKENES AND AROMATICS
edited by Peter Taylor and Michael Gagan

SEPARATION, PURIFICATION AND IDENTIFICATION
edited by Lesley Smart

ELEMENTS OF THE p BLOCK
edited by Charles Harding, David Johnson and Rob Janes

MECHANISM AND SYNTHESIS
edited by Peter Taylor

The Molecular World Course Team

Course Team Chair
Lesley Smart

Open University Authors
Eleanor Crabb (Book 8)
Michael Gagan (Book 3 and Book 7)
Charles Harding (Book 9)
Rob Janes (Book 9)
David Johnson (Book 2, Book 4 and Book 9)
Elaine Moore (Book 6)
Michael Mortimer (Book 5)
Lesley Smart (Book 1, Book 3 and Book 8)
Peter Taylor (Book 5, Book 7 and Book 10)
Judy Thomas (*Study File*)
Ruth Williams (skills, assessment questions)

Other authors whose previous contributions to the earlier courses S246 and S247 have been invaluable in the preparation of this course: Tim Allott, Alan Bassindale, Stuart Bennett, Keith Bolton, John Coyle, John Emsley, Jim Iley, Ray Jones, Joan Mason, Peter Morrod, Jane Nelson, Malcolm Rose, Richard Taylor, Kiki Warr.

Course Manager
Mike Bullivant

Course Team Assistant
Debbie Gingell

Course Editors
Ian Nuttall
Bina Sharma
Peter Twomey

CD-ROM Production
Andrew Bertie
Greg Black
Matthew Brown
Philip Butcher
Chris Denham
Spencer Harben
Peter Mitton
David Palmer

BBC
Rosalind Bain
Stephen Haggard
Melanie Heath
Darren Wycherley
Tim Martin
Jessica Barrington

Course Reader
Cliff Ludman

Course Assessor
Professor Eddie Abel, University of Exeter

Audio and Audiovisual recording
Kirsten Hintner
Andrew Rix

Design
Steve Best
Carl Gibbard
Sarah Hack
Mike Levers
Sian Lewis
John Taylor
Howie Twiner

Library
Judy Thomas

Picture Researchers
Lydia Eaton
Deana Plummer

Technical Assistance
Brandon Cook
Pravin Patel

Consultant Authors
Ronald Dell (*Case Study:* Batteries and Fuel Cells)
Adrian Dobbs (Book 8 and Book 10)
Chris Falshaw (Book 10)
Andrew Galwey (*Case Study:* Acid Rain)
Guy Grant (*Case Study:* Molecular Modelling)
Alan Heaton (*Case Study:* Industrial Organic Chemistry, *Case Study:* Industrial Inorganic Chemistry)
Bob Hill (*Case Study:* Polymers and Gels)
Roger Hill (Book 10)
Anya Hunt (*Case Study:* Forensic Science)
Corrie Imrie (*Case Study:* Liquid Crystals)
Clive McKee (Book 5)
Bob Murray (*Study File*, Book 11)
Andrew Platt (*Case Study:* Forensic Science)
Ray Wallace (*Study File*, Book 11)
Craig Williams (*Case Study:* Zeolites)

CONTENTS

PART 1 CRYSTALS
Lesley Smart

PART 2 MOLECULAR SHAPE

Michael Gagan

CASE STUDY: LIQUID CRYSTALS —
The fourth state of matter

Corrie Imrie

Part 1

Crystals

Lesley Smart

INTRODUCTION

You will be aware how atoms sometimes bond together to make individual molecules, as in carbon dioxide, chlorine and iodine for instance, but how in other cases, like diamond, silica, and sodium chloride, it is not possible to distinguish individual molecules (Figures 1.1a–d). You should also remember that bonding, and thus the arrangement of electrons, influences the shape of a molecule such as ammonia (Figure 1.1e). Many chemicals are solids at normal temperatures, and in the first part of this Book we are going to investigate the variety of structures adopted by elements and compounds in the crystalline solid state. The second part of the Book will concentrate only on the structure and shape of individual molecules.

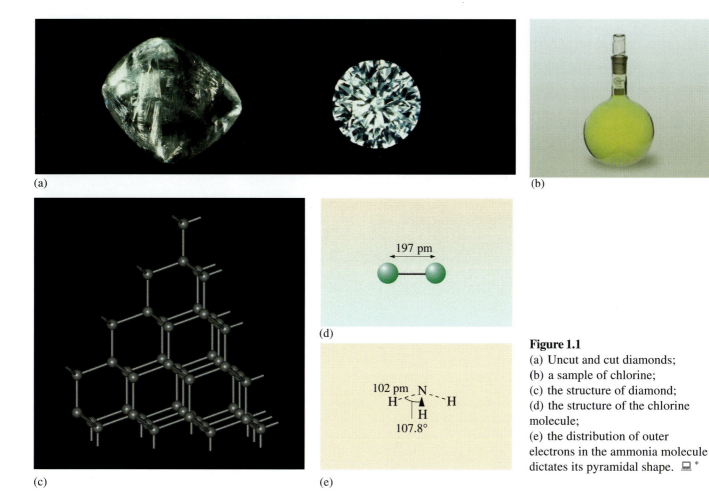

(a)

(b)

(d)

(c)

(e)

Figure 1.1
(a) Uncut and cut diamonds;
(b) a sample of chlorine;
(c) the structure of diamond;
(d) the structure of the chlorine molecule;
(e) the distribution of outer electrons in the ammonia molecule dictates its pyramidal shape. 🖥 *

---

* This symbol, 🖥, indicates that this Figure is available in WebLab ViewerLite™ on one of the CD-ROMs associated with this Book.

To begin with, we are going to look at the structure of metals and of ionic solids. Such materials are very different from those containing individual molecules, in that they comprise extended arrays of atoms or ions in which discrete molecules are not identifiable. For instance, as well as diamond (Figure 1.1c) we can consider another familiar example, sodium chloride (common salt, Figure 1.2a), which has the empirical formula NaCl. Although the formula gives us the highly important information that there is one sodium atom present for every chlorine atom, it disguises the fact that crystalline sodium chloride does not comprise discrete NaCl molecules.

What is the structure of sodium chloride?

It is an aggregate of Na^+ ions and Cl^- ions arranged together in a regular, repeating three-dimensional pattern, with six ions of one type octahedrally surrounding an ion of the other type (Figure 1.2b).

(a)
(b)

Figure 1.2 (a) Crystals of sodium chloride, NaCl; (b) the crystal structure of NaCl.

The structure adopted by NaCl is also adopted by many other compounds, but it is not the only type of crystal structure. Ionic solids adopt a variety of patterns in the arrangements of their constituent ions. We shall show you a few of the ones that are more commonly found, and discuss the reasons for the variations.

We shall then move from ionic solids to look at a variety of structures adopted by solids where covalent rather than ionic bonding is predominant. For instance, silica (quartz), with empirical formula SiO_2 (Figure 1.3), is covalently bonded: in its crystals we find an extended structure of covalently linked silicon and oxygen atoms, but no distinguishable individual molecules of SiO_2.

In the final kind of solid state structure that we consider — molecular structures — small individual, covalently bound molecules such as iodine form crystals in which the molecules are held together by weak bonding (Figure 1.4).

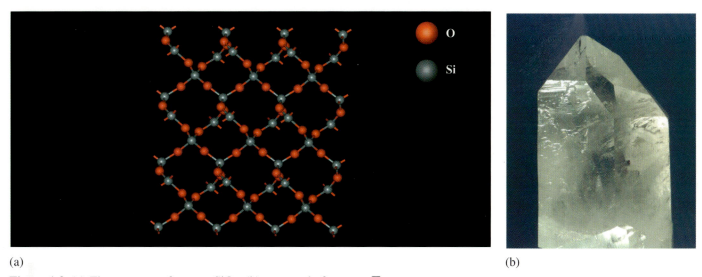

(a)   (b)

Figure 1.3 (a) The structure of quartz, SiO_2; (b) a crystal of quartz. 🖥

(a)

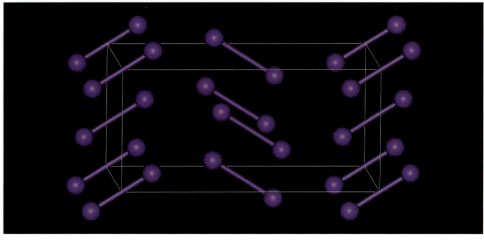

(b)

Figure 1.4
(a) Crystals of iodine;
(b) the structure of solid iodine. 🖥

13

Not everyone finds it an easy matter to visualize things in three dimensions, and so we have devised several ways of demonstrating the structures; you will have to find out which method helps you most. First and foremost is the use of WebLab Viewer Lite™ to look at the structures. Most of the diagrams of crystal structures in this Book can be viewed using WebLab ViewerLite. You will find the files on one of the CD-ROMs associated with this Book; you can access them via the CD-ROM Guide (p. 240) under 'Figures'. WebLab ViewerLite allows you to view a structure in any orientation: the instructions for using it can be found on one of the CD-ROMs*, and if you haven't used it before, you should practise now with the four structures in this introduction. In addition to WebLab ViewerLite on one of the CD-ROMs, there is a program *Virtual Crystals* to work through, model-building exercises to view, and a program entitled *Crystals*, which contains some computer animations. If at all possible, try to study this Book near your computer, so that you can more easily use some or all of these aids.

Many ionic crystal structures can be regarded as based on a few very simple ways of packing spheres together in three dimensions. The simple sphere-packing schemes are best illustrated by the structures of metals, and so it is with these that we begin in the next Section.

---

* Open University students can also access this information from the *Study File*.

STRUCTURES OF METALS

2

2.1 Close-packing in two dimensions

Before moving on to three-dimensional structures, it's convenient first to take a look at the efficiency of close-packing in two dimensions, by looking at two different ways in which circles may be arranged. Figure 2.1a shows a close-packed arrangement, whereas Figure 2.1b contains a square arrangement. The first apparent difference is that each circle touches six others in Figure 2.1a, but only four others in Figure 2.1b. How do the two arrangements compare for efficient use of space? The bounded areas highlighted on the two diagrams both enclose a total of four complete circles. If we overlay these two areas (Figure 2.1c), we can see clearly that the four circles (comprising two whole circles, two half circles, and four quarter circles) in Figure 2.1a occupy less total area than the four circles in Figure 2.1b (comprising one whole circle, four half circles, and four quarter circles) — about 15% less in fact. Now let's move on to the three-dimensional case.

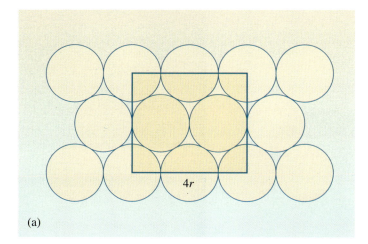

(a)

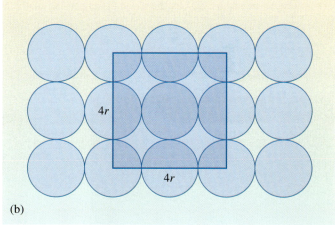

(b)

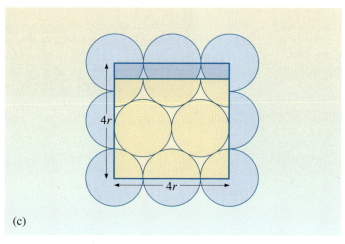

(c)

Figure 2.1 (a) A close-packed layer of circles; (b) a square array of circles; (c) framed area in part (b) overlaid with the framed area of part (a), demonstrating the more efficient use of space in close-packing.

COMPUTER ACTIVITY 2.1 Using computer animation and models to study close-packing

The close-packing of spheres is demonstrated by building models in *Model Building* on one of the CD-ROMs associated with this Book.

There is a computer animation of close-packing in the first part of *Crystals* on the CD-ROM.

Close-packing is also demonstrated on the CD-ROM in the program *Virtual Crystals*.

You should try these activities as soon as possible. Each of them should take you 15–20 minutes to complete.

2.2 Close-packing in three dimensions

A large number of metals adopt what is known as a **close-packed structure** (Figure 2.2). We model these structures by treating the atoms as though they are *hard spheres*. This is a convenient model to use, and modern science shows that it is not unreasonable: Figure 2.3 shows the most detailed pictures of atoms that are currently available, using electron microscopy. In Figure 2.3a we see the image of a single gold atom, and in Figure 2.3b the arrangement of gold atoms on the surface of a gold crystal. Notice the way the gold atoms are arranged on the surface; such an arrangement of atoms makes the most efficient use of the space available.

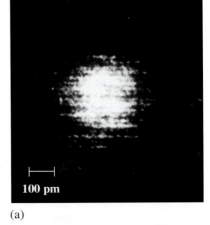

100 pm

(a)

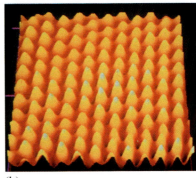

(b)

Figure 2.3
(a) A single atom of gold captured by STEM techniques (scanning transmission electron microscopy);
(b) STM (scanning tunnelling microscopy) of gold atoms on the surface of a crystal of gold.

Figure 2.2 Aluminium, gold and copper all adopt close-packed structures.

We are interested in the structure of solids, which are three-dimensional, and thus in the three-dimensional packing of spheres. We shall concentrate first on a layer of spheres. As with circles, spheres pack together in a layer most efficiently if they are close-packed (Figure 2.4a and b). Looked at from above, the spheres in the close-packed layer have 'curved triangular' spaces between them; for the sake of simplicity we shall refer to them as triangles. These are marked with crosses and dots in Figure 2.4c.

In an infinite layer of spheres, what is the ratio of the number of spaces to the number of spheres?

There are twice as many spaces as there are spheres.

This is easy to check by counting the number of spaces around a sphere in the centre of Figure 2.4c: there are six. Now count how many spheres there are around one space: the answer is three.

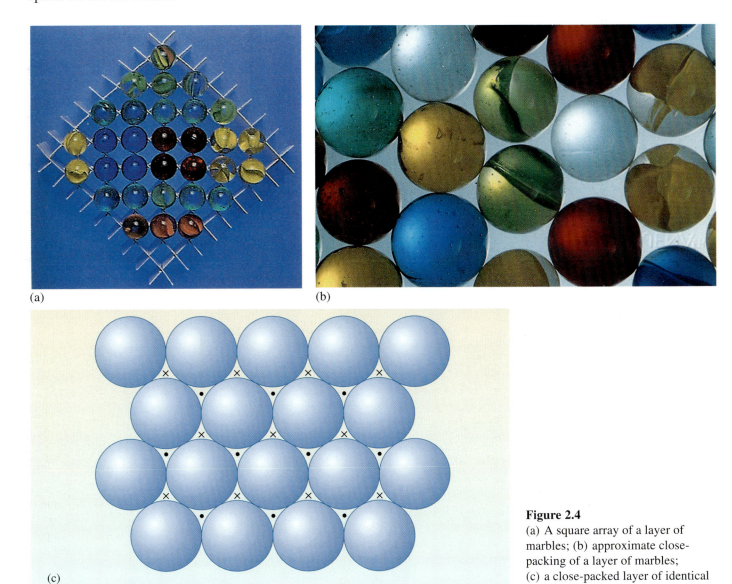

(a)

(b)

(c)

Figure 2.4
(a) A square array of a layer of marbles; (b) approximate close-packing of a layer of marbles; (c) a close-packed layer of identical spheres.

🔵 Do the spaces in Figure 2.4c differ in any way?

⚪ The spaces marked with a cross are characterized by the tips of the triangles pointing up the page. The other triangles, with tips pointing down the page, are marked with dots, and are equal in number to those marked with crosses.

At this stage in our argument these two types are equivalent: Figure 2.4c would look exactly the same if viewed from the other side.

To start to build up the three-dimensional array, a similar close-packed layer is placed on the first. The layers could be stacked so that the spheres in the second layer are placed directly on top of those in the first, but this is not the most efficient use of space. Close-packing is achieved when the spheres of the upper layer are placed over the hollows of the lower layer. In Figure 2.5 the green spheres represent the upper layer (layer B), and the original layer is represented by the blue spheres (layer A). Look carefully at Figure 2.5. The spheres of layer B rest in the hollows or triangular spaces marked with a cross in layer A — that is, in the spaces with the tips pointing toward the top of the diagram. The hollows marked with a spot, with their tips pointing downward, remain uncovered. (We could equally well have used the spheres in layer B to cover the spaces marked with dots, and left uncovered the spaces marked with crosses; it would not have affected the efficiency of the packing.)

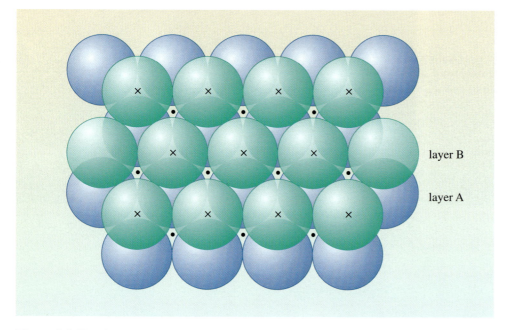

Figure 2.5 Two layers of close-packed spheres.

So far, we have talked of the spaces between the spheres as being curved triangles, but, of course, they are no such thing. Look carefully at Figure 2.6a: the spaces now formed by the layers are of two types. The spaces marked by crosses in Figure 2.6a are bounded by *four* spheres, with their centres arranged at the corners of a tetrahedron. These spaces are known as **tetrahedral holes**[*]; one is shown in a side view as Figure 2.6b. The holes marked by spots in Figure 2.6c are bounded by *six* spheres, with their centres located at the corners of an octahedron: these are called **octahedral holes**[*]; one is shown in a side view as Figure 2.6d.

[*] These are referred to as tetrahedral and octahedral *interstices* on the CD-ROM *Virtual Crystals* program.

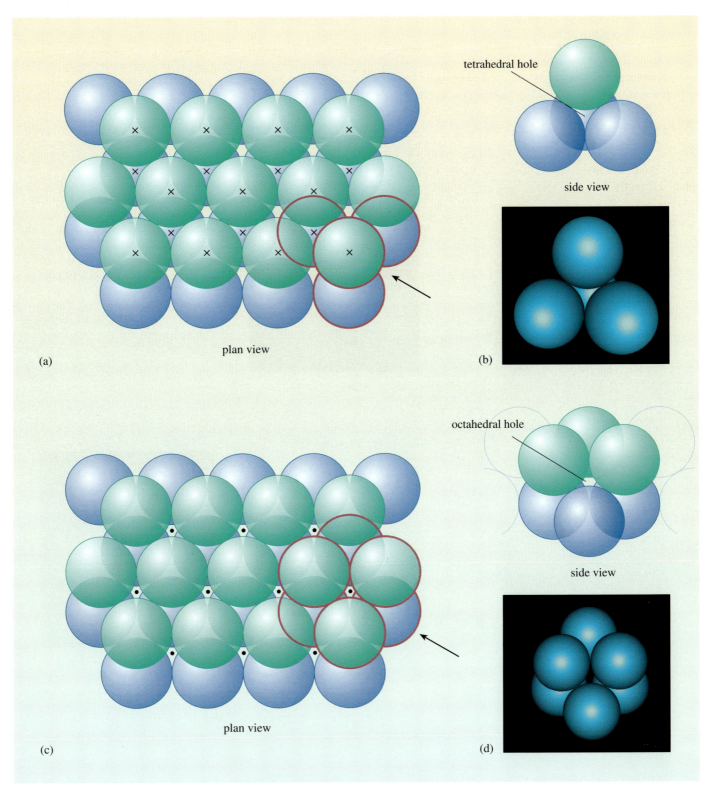

Figure 2.6 (a) Plan view of two layers of close-packed spheres with the tetrahedral holes marked by crosses. (b) Side view (shown as both a line drawing and a WebLab ViewerLite image) of the four spheres outlined in red in part (a) from the direction of the arrow; these spheres enclose a tetrahedral hole in the close-packed array. (c) Plan view of two layers of close-packed spheres with the enclosed octahedral holes marked by spots. (d) Side view (shown as both a line drawing and a WebLab ViewerLite image) of the six spheres outlined in red in part (c); these spheres enclose an octahedral hole in the close-packed array. 🖳

⬤ What do you notice immediately about the relative sizes of the octahedral and tetrahedral holes?

⬤ The octahedral holes are much bigger than the tetrahedral holes. (This is obvious when you think that it takes six spheres to enclose an octahedral hole, and only four to enclose a tetrahedral hole.) It can be calculated that the radius of a sphere that will *just* fit in a tetrahedral hole in a close-packed array of spheres of radius *r*, is 0.225*r* (Figure 2.7a). For an octahedral hole the radius is 0.414*r* (Figure 2.7b).

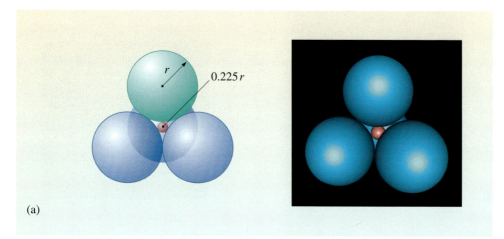

(a)

Figure 2.7
(a) A sphere of radius 0.225*r* just fits into a tetrahedral hole (shown as both a line drawing and a WebLab ViewerLite image);
(b) a sphere of radius 0.414*r* just fits into an octahedral hole (shown as both line drawings in plan view and side view from the arrowed direction, and as a WebLab Viewer Lite image). 🖳

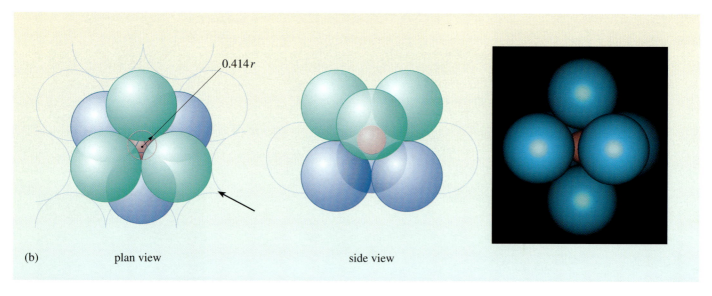

(b)   plan view   side view

⬤ Now look at Figures 2.4 and 2.5 again, and try to work out the proportion of octahedral holes to spheres in the structure.

⬤ From Figure 2.4 you can see that there would be *twice* as many triangles as spheres in an infinite array. When the second layer of spheres is added in Figure 2.5, *half* of these triangles become octahedral holes (the ones with the dots), so there is one octahedral hole for each sphere, and in an infinite three-dimensional array of *n* spheres, there will be *n* octahedral holes.

From Figure 2.6a, you can see that in an infinite array there would be *twice* as many tetrahedral holes as octahedral holes. So, in an infinite three-dimensional array of *n* spheres, there will be 2*n* tetrahedral holes.

So far we have only two layers. When we add a third layer, there is a choice: there are two types of hollow on layer B of Figure 2.5 in which the spheres of the next layer can rest. In the first alternative, the spheres in layer C can be arranged so that they are directly over the spheres in layer A; they are placed over the tetrahedral holes (Figure 2.8). This type of packing is known as **hexagonal close-packing, *hcp***, and is adopted by many of the metallic elements in the Periodic Table, such as Be, Mg, Sc, Ti, Co and Zn, to name but a few (Figure 2.9). The layers in a hexagonal close-packed structure are designated ABAB... . Notice that in hexagonal close-packing the positions marked with a dot are always occupied by spaces and never by spheres. Thus, in models of this type of packing, a thin wire could be drawn vertically through the layers using these narrow channels.

Figure 2.9
Zinc metal adopts the *hcp* structure.

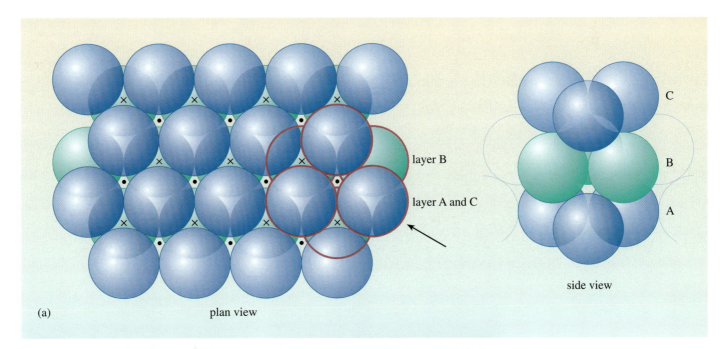

(a) plan view — layer B, layer A and C — side view

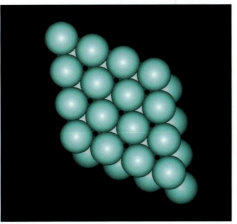

(b)

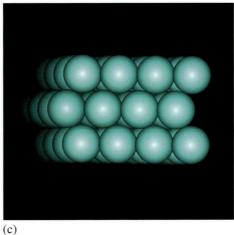

(c)

Figure 2.8
(a) Plan view and side view (of spheres outlined in red from arrowed direction) of three layers of close-packed spheres in a hexagonal close-packed, *hcp*, arrangement; (b) computer representation of three *hcp* layers, showing the narrow channels; (c) three *hcp* layers showing the ABAB... stacking sequence.

The alternative arrangement for a third layer on Figure 2.5, is for the layer C spheres not to lie directly over the spheres of layer A, but over the spaces marked with spots — that is, over the octahedral holes (Figure 2.10). This is known as **cubic close-packing,** *ccp*, and is designated ABCABC... . An easy way to remember the difference between hexagonal and cubic close-packing, is that both the word 'cubic' and the sequence ABC, end with the letter 'c'.

This cubic close-packed structure is also adopted by many of the metallic elements, of which gold is one example. The close-packed planes of atoms are to be found parallel to the body-diagonal in the cubes (Figures 2.11a and b), and one of these planes on the surface of a gold sample has been captured dramatically by an STM picture (Figure 2.3b).

STUDY NOTE

Remember that many of the structures in this Book — look for this symbol 💻 — can be viewed using WebLab ViewerLite. You will find them in the 'Figures' folder on one of the CD-ROMs associated with this Book; first go to the CD-ROM Guide. Instructions on how to use WebLab ViewerLite are on the CD-ROM.

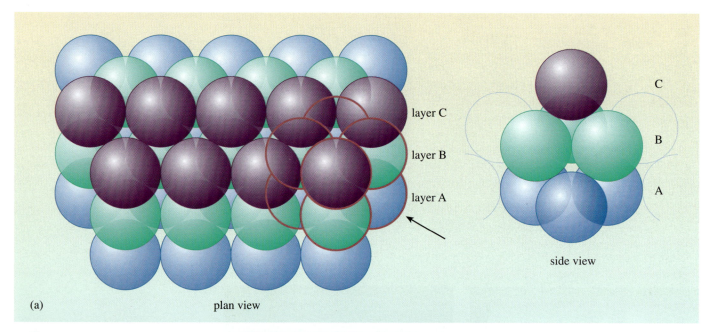

layer C
layer B
layer A

side view

C
B
A

(a)   plan view

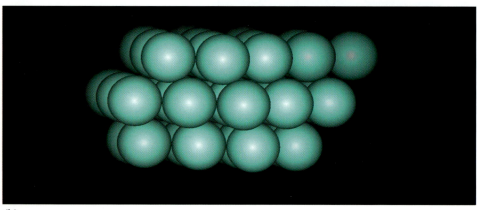

(b)

Figure 2.10
(a) Three layers of close-packed spheres in a cubic close-packed, *ccp*, arrangement, shown in plan view and side view (of spheres outlined in red from arrowed direction); (b) computer representation of three *ccp* layers. 💻

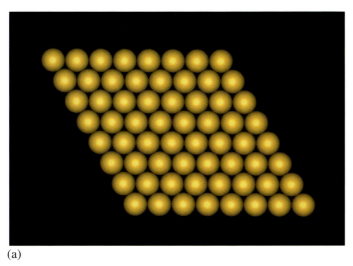
(a)

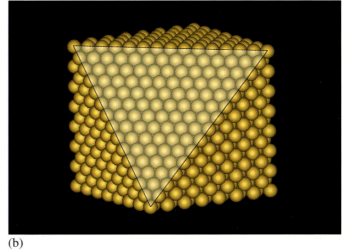
(b)

Figure 2.11 The structure of gold: (a) close-packed layer; (b) position of close-packed layer (toned triangle) in a cubic crystal. 💻

BOX 2.1 Crystals in action 1: colloidal gold

Colloidal gold is formed by striking an electric arc between two gold wires under water, or by reducing aqueous solutions of gold trichloride, $AuCl_3$, with various reducing agents such as iron sulphate, $FeSO_4$. Michael Faraday made the discovery that the tiny particles thus formed are minute crystals of gold, and according to the sizes and shapes of these particles, the colloidal solution has different colours: larger particles give blue, and smaller ones red (Figure 2.12a).

If the $AuCl_3$ is reduced with a mixture of tin dichloride and tin tetrachlorides ($SnCl_2$ and $SnCl_4$, respectively) a purple powder precipitates, which is thought to be tin oxide with colloidal gold adsorbed on the surface. This powder is used to make ruby glass (Figure 2.12b), the colour being due to the minute crystalline particles of gold. The Romans made gold ruby glass (there is a fine example known as the Lycurgus Cup in the British Museum), but the secret of how to do it was lost for many years and was only rediscovered in the seventeenth century by Andreas Cassius in Bohemia, whose name is remembered in the powder —*purple of Cassius*.

Figure 2.12
(a) Faraday's original solution of colloidal gold; (b) ruby glass.

(a)

(b)

There are innumerable stacking sequences possible that may be described as close-packed, where the repeating sequence of layers is not as simple as in the two cases described above. The hexagonal close-packed sequence (ABABAB…) and the cubic close-packed sequence (ABCABC…) are the two simplest sequences, and are those most commonly encountered in the crystal structures of the metallic elements.

Now look carefully at Figures 2.8a and 2.10a again. Select a sphere in the middle of layer B, and in each case count the total number of spheres actually touching it. (You can do this in WebLab ViewerLite by clicking on a particular atom to select it.)

● How many spheres did you count?

● You should have counted twelve: in both cases there are six in the second layer, three in the plane above and three in the plane below.

This number of *equidistant nearest neighbours* is known as the **coordination number**. You will encounter this term many times in chemistry.

STUDY NOTE

Remember the three programs on the CD-ROM which are available to assist your understanding of close-packing: *Model Building*, *Crystals* and *Virtual Crystals*.

A set of ten interactive self-assessment questions is provided on *The Third Dimension* CD-ROM A. The questions are scored, and you can come back to the questions as many or as few times as you wish in order to improve your score on some or all of them. This is a good way of reinforcing the knowledge you have gained while studying this Book.

QUESTION 2.1

How many tetrahedral holes and how many octahedral holes are there in a *ccp* array of *n* spheres?

QUESTION 2.2

What is the coordination number of any atom in an infinite *ccp* array?

2.3 Body-centred and primitive cubic structures

Some metals, such as the alkali metals of Group 1 (Figure 2.13), adopt a structure that is somewhat less efficiently packed than *hcp* or *ccp*. This is the **body-centred cubic (*bcc*)** structure, in which an atom in the centre of a cube is surrounded by *eight* equidistant nearest neighbours, situated at the corners of the cube. A portion of a body-centred cubic array is shown in Figure 2.14a. You should be able to see that the cube formed by eight spheres contains a sphere occupying the centre of the cube, thus giving the structure its name.

It is common in structures for the positions of the atoms to be represented by small spheres that do not touch (Figure 2.14 b and c); such diagrams are commonly known as **ball-and-stick models**: this is merely a device to open up the structure and allow it to be seen more clearly, even though it is a less realistic representation of a real crystal than the **space-filling model** we used in Figure 2.14a. We shall discuss the relative sizes of atoms and ions in Section 5.

Figure 2.13
Sodium metal (kept under oil, as it reacts with the atmosphere).

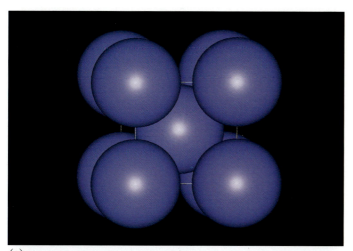

(a)

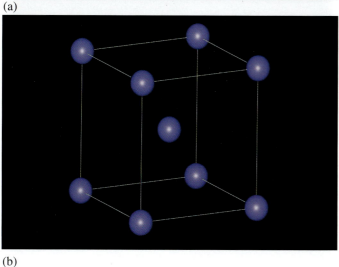

(b)

Figure 2.14 A body-centred cubic array, as adopted by sodium, potassium, etc.: (a) one cube, space-filling representation; (b) one cube, ball-and-stick model; (c) four cubes, ball-and-stick model.

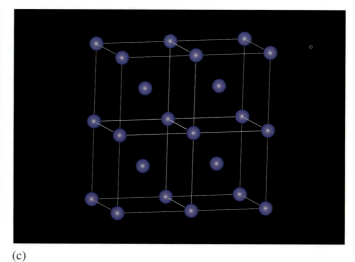

(c)

In the *bcc* structure, each atom is surrounded by eight atoms at the corners of the cube, but it has six other atoms only slightly more remote at the body-centres of the adjacent cubes (Figure 2.15). This means that the packing efficiency in *bcc* is not very different from that in close-packed structures: 68% of the volume is occupied, compared with the 74% of volume occupied in close-packing, regardless of the atom sizes.

QUESTION 2.3

If the length of a side of the cube in Figure 2.14b is a, what is the shortest distance from the atom at the cube centre to (a) an atom at a corner of the cube, and (b) the equivalent position in an adjacent cube? (See the Maths Help overleaf.)

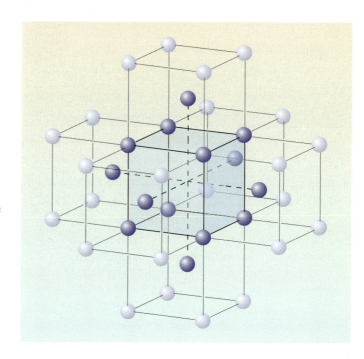

Figure 2.15 Body-centred cube, showing the six next-nearest neighbours at the body-centres of the adjoining cubes.

MATHS HELP: THE PYTHAGORAS THEOREM

This states that for any right-angled triangle, the square on the hypotenuse (side facing the right-angle) is equal to the sum of the squares on the other two sides. Thus, for the triangle shown in Figure 2.16:

$$a^2 + b^2 = c^2$$

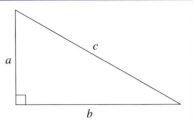

Figure 2.16
A right-angled triangle with sides a and b, and hypotenuse, c.

QUESTION 2.4

In a right-angled triangle with a hypotenuse of 5 cm, and one side of 3 cm, what is the length of the third side?

STUDY NOTE

More maths help is available in *The Sciences Good Study Guide*[1] (see the Further Reading list on p.118).

Another cubic structure, and the simplest, is the so-called **primitive cubic structure**. There is only one metal that adopts this structure — polonium.

Figure 2.17a shows how a primitive cubic structure is built from spheres by forming a square array, with a second layer directly over the first layer, and so on. Figure 2.17b uses a ball-and-stick representation to show how each atom in such a structure sits at the corner of a cube.

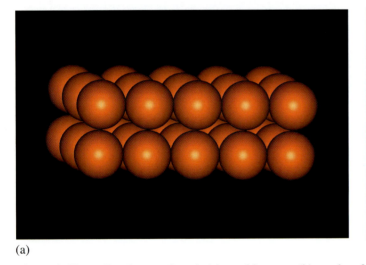

(a)

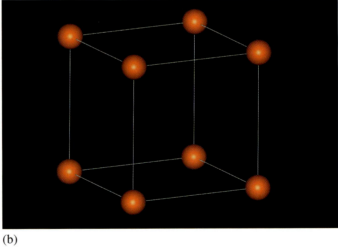

(b)

Figure 2.17 (a) Two layers of a primitive cubic array; (b) a cube of atoms from a primitive cubic array.

MODEL EXERCISE 2.1
Build a model of a primitive cubic structure

You will spend much time making molecular models in Part 2 *Molecular Shape*. The rationale for using models is explained in Box 1.2 of Part 2, p. 125. We have based our model diagrams like Figure 2.18 on the Orbit kit, but the principles involved are the same in other model systems.

Use your model kit to build a model of the primitive cubic structure, with octahedral atom centres (Figure 2.18).

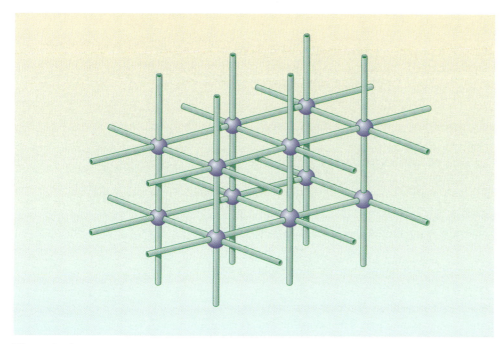

Figure 2.18 Model of a primitive cubic structure.

Most of the metallic elements have one of the three structures, *hcp*, *ccp* or *bcc* (although a few of the lanthanides and actinides have a mixed *hcp/ccp* structure). You can gain some impression of the overall distribution of the stable metal structures at 25 °C from the Periodic Table in Figure 2.19; there doesn't appear to be any obvious pattern as to which structure is adopted, although there is a tendency for elements in the same Group to adopt the same structure.

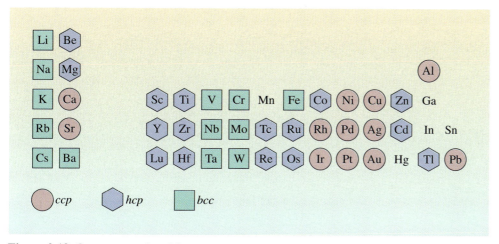

Figure 2.19 Occurrence of packing types among the metals. Note that some metals (Mn, Ga, In and Sn) do not adopt these regular structures.

2.4 Summary of Section 2

1   In two dimensions, a close-packed array of circles is approximately 15% more efficient in its use of space than a square array of circles.

2   In a close-packed array of spheres, there are two different types of curved space between the spheres.

3   In an infinite layer of close-packed spheres there are twice as many spaces between the spheres as there are spheres.

4   In a three-dimensional close-packed array of spheres, two arrangements are common, ABABAB…, known as hexagonal close-packing, *hcp*, and ABCABCABC…, known as cubic close-packing, *ccp*. These commonly occur in simple crystal structures such as those of the metallic elements.

5   In a three-dimensional close-packed array of n spheres, there are $2n$ tetrahedral holes and n octahedral holes.

6   The coordination number (closest number of equidistant nearest neighbours) of any atom in a close-packed array is twelve.

7   A body-centred cubic structure consists of atoms at each corner of a cube surrounding one identical atom in the centre of the cube. The coordination number of any atom in this array is eight.

8   A primitive cubic array only has atoms at each corner of a cube.

QUESTION 2.5

What is the coordination number of an atom in an infinite primitive cubic array? What geometrical arrangement is adopted by the nearest coordinating atoms?

QUESTION 2.6

What is the coordination of an atom occupying (a) a tetrahedral hole, (b) an octahedral hole?

THE INTERNAL STRUCTURE OF A CRYSTAL

3

3.1 Lattice points and lattices

Most chemicals are solid at room temperature, and of these, most are in the crystalline state. The most important feature of crystals is that they possess a very high degree of internal order; that is, the basic components of which the crystal is composed are arranged in a perfectly regular way, repeating over and over again in all directions. The repeating units within a crystal may be atoms (for instance, in the crystalline metals that we looked at in Section 2), ions (e.g. sodium chloride), or covalent molecules (e.g. iodine).

Crystals usually have flat shiny faces bounded by straight edges (Figure 3.1). In 1664, Robert Hooke (see Box 3.1 and Figure 3.2) noted the regular form of crystals, and suggested that this resulted from ordered packing within the crystal. In 1671, Niels Stensen (see Box 3.2 and Figure 3.4) observed that, although crystals of the same substance apparently vary considerably in shape, this is only because some faces develop more than others, and that if you measure the angles between similar faces on different crystals they are always constant (Figure 3.3). This constancy of interfacial angles reflects the internal order within the crystal.

Figure 3.1 From left to right, crystals of mica, calcite, pyroxene, pyrite and quartz showing flat, shiny faces.

(a)

(b)

Figure 3.2 (a) Willen Church. A small but very beautiful seventeenth-century church at Willen in Milton Keynes, designed by Robert Hooke (1635–1703); (b) stained-glass window to the memory of Robert Hooke, St Helen's, Bishopgate (bottom row, second from right).

(b)

(a)

Figure 3.3 Crystals of (a) quartz, and (b) pyrite, exhibiting the constancy of interfacial angles.

BOX 3.1 Robert Hooke 1635–1703

Born in Freshwater, Isle of Wight, Robert Hooke was an amazingly gifted man. He was a scientist, engineer and architect, and a founding member of the Royal Society (of which he was curator for 50 years). Apart from the law of elasticity that bears his name, he also invented the air pump, the universal joint, the cross filaments of the telescope, and he developed the balance spring for watches, thus paving the way for the Huygens working model. He discovered the red spot on Jupiter, which was named 'Hooke's spot' in his honour. He published the first book on microscopy, his famous *Micrographia*, of which Samuel Pepys said in his diary, 'I sat up till 2 o'clock in my chamber, reading of Mr Hook's *Microscopical Observations*, the most ingenious book that ever I read in my life.'.

He was friend and assistant to Robert Boyle, and undoubtedly helped him to formulate Boyle's law. It was also his insight on an inverse-square force law in elliptical orbits that led Newton to his universal law of gravitation: Newton's lack of acknowledgement of his contributions led to a lifelong rift between them.

Hooke was also a talented architect and was a close friend of Sir Christopher Wren. Together, they helped redesign and rebuild London after the Great Fire of 1666, Hooke being responsible for the new Bethlehem Hospital (the Bedlam), the Royal College of Physicians (both demolished in the nineteenth century) and the Monument. Two other examples of his talents remain, Ragley Hall in Warwickshire and Willen Church in Milton Keynes (Figure 3.2a).

As a lecturer at Gresham College, he was not permitted to marry, but managed to maintain a rich personal life, with a succession of housekeepers, all of which is detailed in his diaries. Referring to his final housekeeper, his niece Grace, he comments 'Grace perfecte intime omne. Slept well.'.

No portraits of Hooke have survived, it being suggested that Newton destroyed the one at the Royal Society, and the stained-glass window erected to his memory (Figure 3.2b) was lost in the IRA bombing of the City of London in 1996.

BOX 3.2 Niels Stensen 1638–1686

Born in Copenhagen, Niels Stensen (Figure 3.4) devoted his early life to medical research, discovering the Steno duct of the parotid gland in the ear, and the function of the ovaries.

After his conversion to Roman Catholicism on All Souls Day 1667, he lived much of his time in Florence, and eventually entered the priesthood, becoming a bishop in 1677.

His later research lay in palaeontology, geology and mineralogy, and he was the first to point out the true origin of fossil animals, and to differentiate sedimentary and volcanic rocks. His great work, the *Prodomus*, outlines the principles of modern geology.

Figure 3.4
Niels Stensen (also known as Nicolaus Steno, 1638–1686).

Crystalline solids consist of regular arrays of atoms, ions or molecules, where the distances between the atoms are of the order of 100 pm (1×10^{-10} m). If you replace positions in a crystal which have *identical surroundings*, with a dot or a point, then these dots are called **lattice points**, and the array of lattice points is defined as the **crystal lattice**. For simplicity, we have shown this in two dimensions in Figure 3.5a and b, where an array of 'molecules' in (a) is replaced by lattice points in (b). Figure 3.5c shows part of a simple cubic three-dimensional lattice.

(a)

(b)

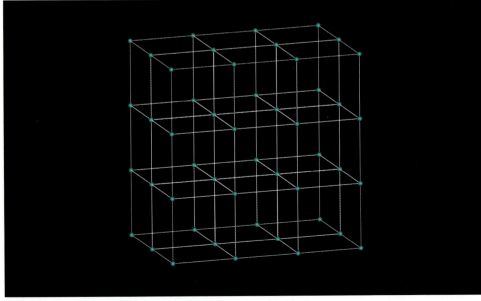

(c)

Figure 3.5
(a) 2-D array of objects;
(b) 2-D lattice of the array in (a);
(c) a simple cubic 3-D lattice. 🖥

Because the structure is regular, or 'periodic', a crystal can act as a three-dimensional diffraction grating to radiation of suitable wavelength; this is illustrated in two dimensions in Figure 3.6. For diffraction to occur, the wavelength of the radiation must be of a similar magnitude to the spacings of the grating.

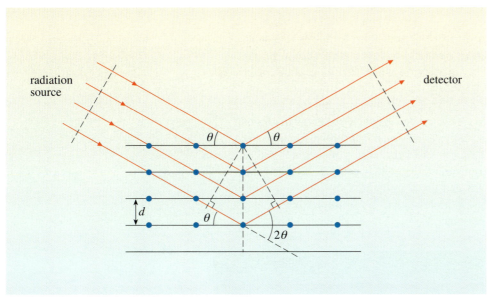

Figure 3.6 Diffraction of electromagnetic radiation incident at angle θ to crystal planes distance d apart.

● What wavelength would you expect to be suitable for diffraction by a crystal, and in what region of the electromagnetic spectrum does this fall (use the *Data Book* from the CD-ROM if necessary)?

● Interatomic distances in a crystal are around 100–200 pm; this wavelength falls in the X-ray region of the electromagnetic spectrum.

As you might expect, the **X-ray diffraction** patterns obtained from crystals are extremely complicated. However, in most cases the position and intensity of the diffracted peaks can be interpreted to give a complete analysis of the internal structure of the crystal, and to provide accurate values of the bond lengths and bond angles. X-ray crystallography started by elucidating relatively simple ionic structures, capturing the diffraction data on film and performing the lengthy calculations by hand. Much of the early work on ionic crystals, such as sodium chloride, was carried out by William and Lawrence Bragg (father and son; see Box 3.3). Today, with sophisticated diffractometers for collecting the data and fast computers for analysis, the technique of **X-ray crystallography** for the determination of crystal structures containing small molecules is almost routine — so much so, that nowadays it is often used by synthetic chemists as an analytical technique for determining the nature of a new product. X-ray crystallography collects diffraction data from small, perfect, so-called 'single crystals'. One of the limiting factors to the use of the technique can be the ability to prepare a good quality crystal, and it has sometimes become possible in recent years, to determine structures by collecting X-ray data from crystalline powders, using a technique called *Rietveld refinement* based on the shapes of the lines in the diffraction pattern. Figure 3.10 shows Dorothy Crowfoot Hodgkin (Box 3.4, p. 37), who, with J. D. Bernal at Cambridge, started using X-ray diffraction to look at biological systems and determined the structure of one of the first biologically important compounds, penicillin (Figure 3.7). The problems of solving the structures of molecules in living systems were immense, because of their sheer size. In Figure 3.8a we see Maurice Wilkins, Francis Crick and James Watson, whose work on the structure determination of DNA was rewarded by the Nobel Prize for Medicine in 1962, and John Kendrew and Max Perutz, who received the Nobel Prize for Chemistry for the structure determination of haemoglobin, both groups using X-ray diffraction (an X-ray photograph of DNA is shown in Figure 3.8b, and Figure 3.8d shows a model of its structure). Sadly, Rosalind Franklin, Figure 3.8c, who did much of the pioneering crystallography of DNA, was not named in this citation, as she died before the award was made, and Nobel Prizes are not awarded posthumously. X-ray crystallographic research now extends into the determination of the structures of very large molecules in biological systems. Improved data collection, using sophisticated diffractometers and vastly improved computing capabilities, means that the complicated structures of proteins and enzymes are frequently solved; the example shown in Figure 3.9 is an enzyme important in research into antibiotic resistance.

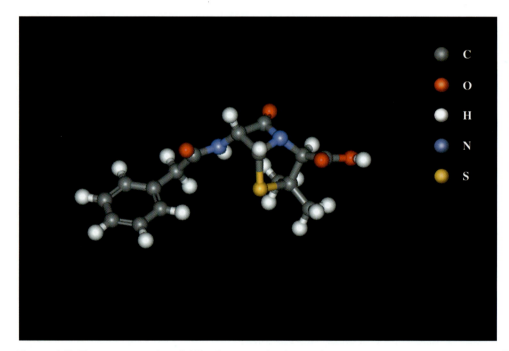

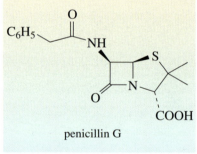

penicillin G

Figure 3.7 The structure of penicillin G (WebLab ViewerLite image and structural formula).

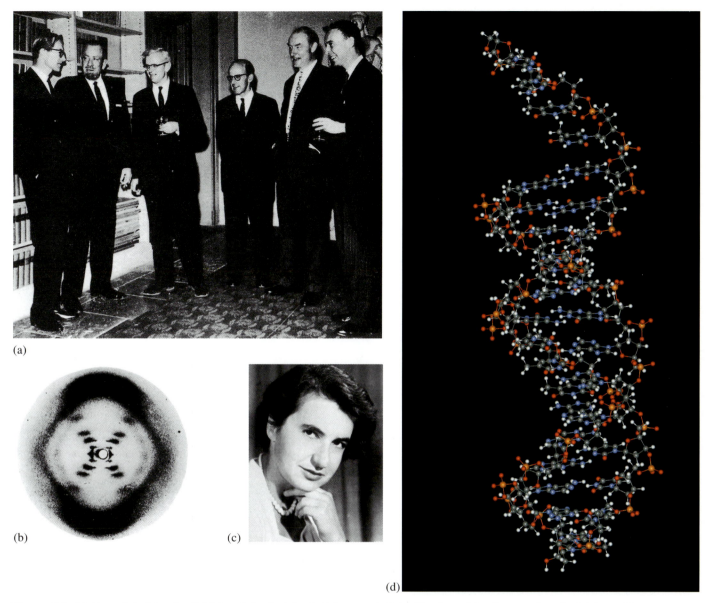

Figure 3.8 (a) From left, Maurice Wilkins, John Steinbeck (Literature), John Kendrew, Max Perutz, Francis Crick, and James Watson in Stockholm in 1962 to receive their Nobel Prizes; (b) an X-ray photograph of DNA in the B-form, taken by Rosalind Franklin late in 1952; (c) Rosalind Franklin; (d) the structure of DNA. ⌨

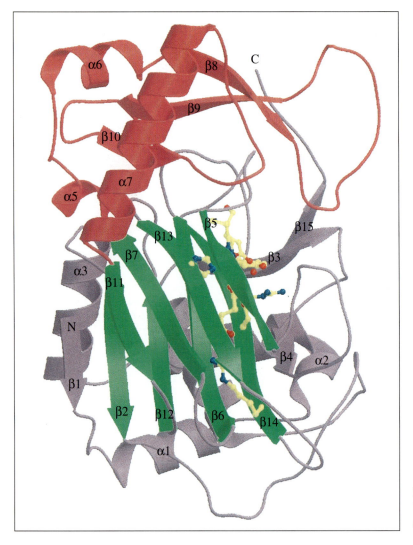

Figure 3.9
The structure of the enzyme clavaminic acid synthase as determined by X-ray crystallography.

BOX 3.3 William and Lawrence Bragg

After a brilliant career at Cambridge, William Henry Bragg took a chair at Adelaide, Australia, at the age of 23. For the next 18 years he concentrated on teaching, administration and the outdoor life, and he published little. In 1904, he was spurred back into active research by some work of Ernest Rutherford on ionizing radiations, and started publishing vigorously. He returned to England in 1908 to take a professorship at the University of Leeds.

In 1912, Friedrich, Knipping and M. von Laue observed the diffraction of X-rays by a crystal. Bragg's son, Lawrence, showed how this could be regarded as a reflection of the X-rays from the planes in the crystal (Figure 3.6), and derived the famous Bragg equation ($n\lambda = 2d \sin \theta$), which connects the wavelength of the X-rays, λ, the perpendicular spacing between the planes, d, and the angle of reflection, θ; n is an integer.

The two Braggs then began a period of collaboration that produced their most brilliant work. They elucidated the crystal structures of many compounds, including ZnS, NaCl and diamond. In 1915, they were jointly awarded the Nobel Prize for Chemistry; Lawrence, aged 25, was the youngest person ever to receive the Prize.

BOX 3.4 Dorothy Crowfoot Hodgkin 1910–1994

Dorothy Hodgkin (Figure 3.10) was born in 1910 in Cairo, Egypt. She went to school in Beccles, Suffolk, and then to Somerville College, Oxford. After graduation, she went to Cambridge and studied crystallography with J. D. Bernal, but shortly returned to Oxford, where she worked for the next thirty-three years. She is one of many women who have excelled in the field of crystallography.

She first elucidated the structure of penicillin G (Figure 3.7), and then turned her attention to vitamin B12, used in the treatment of pernicious anaemia. Her success in determining this structure won her the Nobel Prize for Chemistry in 1964. The following year, she was appointed to the Order of Merit, the first woman to be given this honour since Florence Nightingale. In 1969, she finally solved the structure of the hormone insulin. She retired in 1988 after eighteen years as the Chancellor of Bristol University.

Figure 3.10
Dorothy Crowfoot Hodgkin (1910–1994)

3.2 The unit cell

A crystal lattice can be regarded as an assembly of repeating, identical building blocks — representative of the entire structure — which stack directly on top of each other. Each individual block has a lattice point at each corner. Remember that the lattice points denote positions in the structure with *identical* environments. The representative repeating block is known as the **unit cell**.

Once again, for simplicity, we start by considering the two-dimensional case: Figure 3.11a shows the two-dimensional array from Figure 3.5a, on which are now marked several different unit cells; repeating *any* of these units, edge-to-edge, would reproduce the array. Each unit cell is a parallelogram (an area contained between two pairs of intersecting parallel lines). These may be drawn at any angle, provided that they pass through equivalent positions in the pattern. So for a two-dimensional pattern, a unit cell may be *any* parallelogram, provided that its four corners are equivalent points in the pattern.

There is not just one correct unit cell that can be chosen for a particular lattice; there are always many possible choices, and you simply select a convenient one. In Figure 3.11a, cell **2** is one of the smallest unit cells possible, with lattice points only at the corners (as with cells **3** and **4**); it is known as a **primitive unit cell**. This unit cell would probably not be the best choice, as it does not show the very obvious rectangular shape of the lattice. Two other unit cells are marked **1a** and **1b**; these have the same area because they are parallelograms on the same base, and have the same height. The conventional cell to choose in this example would be unit cell **1a**, as it is most representative of the shape of the lattice, with a lattice point at the centre, reflecting the particular symmetry of this lattice; it is called a **centred unit cell**. The unit cells must fit together to fill all space, and this is illustrated with some tiling patterns in Figure 3.11b. Figure 3.11c shows that not all shapes fit together in a regular repeating pattern to fill the whole space.

In crystals, unit cells are three dimensional, and must also have space-filling proper-ties; they are the building blocks of the structure, but now they stack directly on one

(a)

(b)

(c)

Figure 3.11
(a) 2-D array showing a selection of possible unit cells; (b) tiling: equilateral triangles, hexagons and parallelograms are space-filling; (c) circles, pentagons, and regular polygons beyond hexagons, are not.

another through the faces to form the structure. A general **unit cell** in three dimensions is a parallelepiped* (the 3-D equivalent of a parallelogram) defined by three repeat distances a, b and c, and three angles α, β, and γ (see Figure 3.12).

Since unit cells must pack together to fill all space, there is a limit to the number of possible shapes for the unit cell. The unit cell can take seven possible space-filling shapes. These are illustrated in Figure 3.13a.

- If you think of a unit cell based on a cube, what can you say about its dimensions?

- For it to be an exact cube, $a = b = c$ and $\alpha = \beta = \gamma = 90°$.

Crystals with a **cubic unit cell** are the ones that we shall come across most often in this Book, and in Figure 3.13b, we see how the unit cells stack together to form a three-dimensional crystal. You will also meet tetragonal crystals, which have a **tetragonal unit cell** in which all the angles are 90°, and two of the sides are the same ($a = b$) but different from the third (c), and orthorhombic crystals, which have an **orthorhombic unit cell** in which all the angles are 90° and all three sides differ in length.

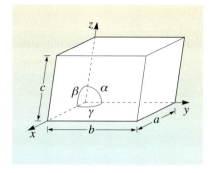

Figure 3.12
Definition of unit cell dimensions and angles for a general unit cell.

* Pronounced '*parallel-a-pie-ped*'.

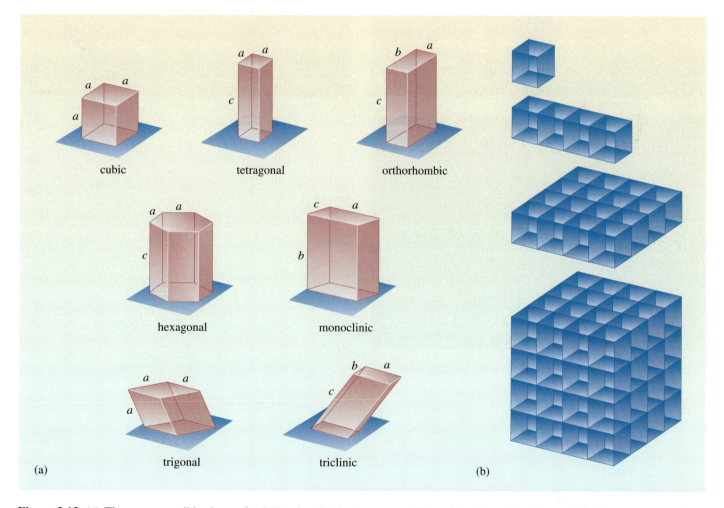

cubic

tetragonal

orthorhombic

hexagonal

monoclinic

trigonal

triclinic

(a)

(b)

Figure 3.13 (a) The seven possible shapes for 3-D unit cells that are space-filling; (b) cubic unit cells assembled in one, two, and three dimensions.

We noted that for a two-dimensional lattice it was possible to have a primitive or a centred lattice; in three dimensions there are more possibilities. The **primitive lattice** is given the symbol **P**; as before, this **primitive unit cell** has a lattice point at each corner only. In this Book, we shall also meet two types of centred lattice:

- **body-centred lattice**, symbol **I**, which has a lattice point at each corner and one in the centre of the cell.
- **face-centred lattice**, symbol **F**, which has a lattice point at each corner and one in the centre of each face of the unit cell.

The corresponding unit cells are illustrated in Figure 3.14.

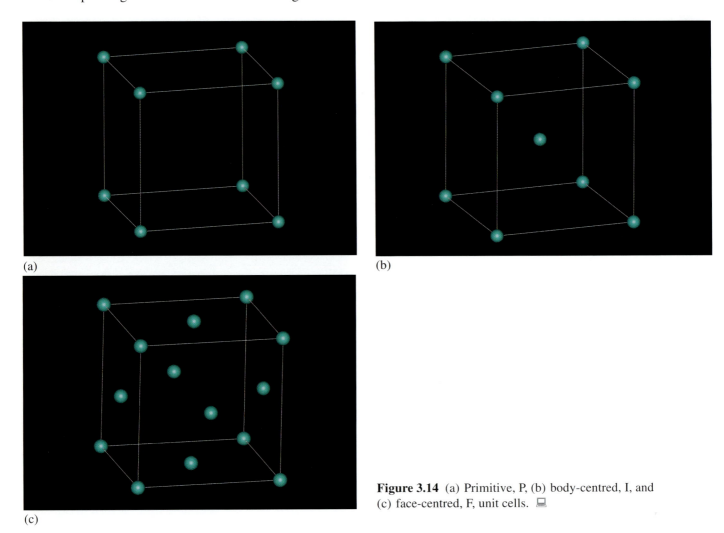

(a)

(b)

(c)

Figure 3.14 (a) Primitive, P, (b) body-centred, I, and (c) face-centred, F, unit cells. 🖥

Crystals are actually made up of repeating arrays of atoms, ions or molecules, and where we have used a spot to indicate a lattice point, this may be occupied in reality by an atom, an ion, a part of a molecule, a whole molecule or even a group of molecules. The lattice is constructed in our imagination by placing spots at identical sites in the structure, and it is used merely to simplify the repeating patterns within a structure. It tells us nothing of the chemistry or bonding within the crystal; for that, we have to know the atomic positions and distances, and we shall see plenty of examples of real structures a little later in this Book.

If we assume the simplest situation, namely that each lattice point is occupied by an atom, we can work out how many atoms there are in a particular unit cell. The number of unit cells sharing a particular atom depends on where the atom is sitting in the cell. We can illustrate this simply, using a two-dimensional square array (Figure 3.15). (The same principles would apply if we assumed that a molecule was sited at each lattice point.)

Figure 3.15
A square array of atoms lying in a plane at the surface of a crystal.

In what types of crystal structure would we find a plane of atoms that looks like this?

A primitive cubic array of atoms would have planes that look like this; polonium is the example that was given earlier. Planes parallel to the cube faces in *bcc* metal structures (Na, K, etc.) would also have this appearance.

Now draw the repeat unit cell for this structure, and calculate the number of atoms in the cell.

You should have drawn the cell shown in Figure 3.16a, repeated to reproduce the plane of atoms in Figure 3.16b: it has a quarter of an atom at each corner. Overall, therefore, the unit cell contains only *one* atom.

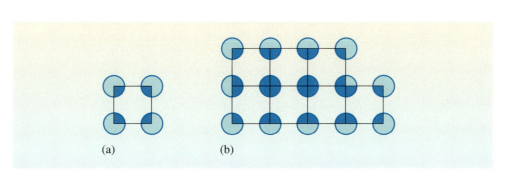

(a)   (b)

Figure 3.16
(a) The unit cell of the array in Figure 3.15, showing the quarter atoms in dark blue; (b) the array generated by repeating the unit cells.

In a three-dimensional unit cell, an atom or molecule at the body-centre of a cell is clearly not shared by any other unit cell (cf. Figure 2.14b). However, this is not true for atoms/molecules at the corners, on an edge, or in a face of a unit cell.

Try to work out how many unit cells share an atom that is (a) in the centre of a face, (b) on a unit cell edge, and (c) at a unit cell corner.

Figure 3.17 illustrates the answers: (a) an atom at the centre of a face is shared between two unit cells; (b) an atom on an edge is shared between four unit cells; (c) an atom at a corner site is shared between eight unit cells.

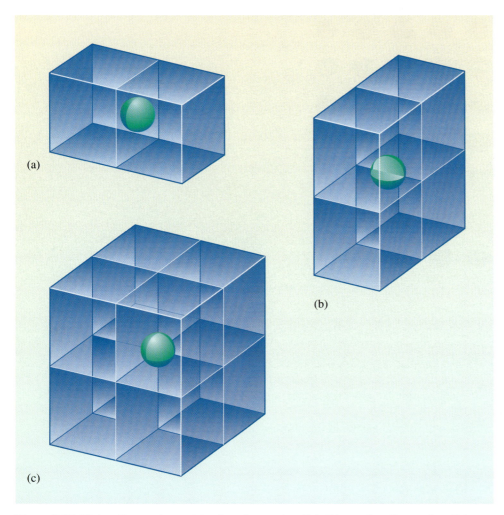

Figure 3.17 Unit cells meeting at (a) a face (two unit cells), (b) an edge (four unit cells), and (c) a corner (eight unit cells).

QUESTION 3.1

Assume that identical atoms are placed at each lattice point for each of the unit cells in Figure 3.14 (P, I and F). Count up the number of atoms in each cell, and see if you agree with our results in Table 3.1, p. 44. (For comments on how the values in Table 3.1 are obtained, see the answer to Question 3.1 on p. 110.)

CALCULATING THE DENSITY OF A CRYSTAL

If we know both the size of the unit cell and what atoms it contains, then we are able to compute the density of a crystal. Take sodium as an example. The unit cell is body-centred, and it therefore contains two atoms of sodium (see Table 3.1, p. 44). The relative atomic mass of sodium is 22.989 8. The unit cell dimension is 428.56 pm.

To determine the mass of sodium in a single unit cell we need to use Avogadro's number. We know that the mass of 1 mole of sodium is 22.989 8 g, and that this contains an Avogadro number of Na atoms.

So, 22.989 8 g of Na contain $6.022\,0 \times 10^{23}$ atoms; two atoms of Na therefore have a mass of

$$\frac{2 \times 22.989\,8\,g}{6.022\,0 \times 10^{23}} = 7.635\,3 \times 10^{-23}\,g$$

Traditionally, density has been quoted in grams per cm^3 ($g\,cm^{-3}$). As these values are more physically meaningful, we will work this value out first of all, before converting to the strict SI unit of $kg\,m^{-3}$.

The volume of the unit cell is $(428.56\,pm)^3 = (428.56 \times 10^{-10}\,cm)^3$
$$= 7.871\,1 \times 10^{-23}\,cm^3$$

$$\text{density} = \text{mass/volume} = \frac{7.635\,3 \times 10^{-23}\,g}{7.871\,1 \times 10^{-23}\,cm^3} = 0.970\,04\,g\,cm^{-3}$$

This makes the density of sodium just slightly less than that of water, which is approximately $1\,g\,cm^{-3}$.

Expressed as an SI unit, this is $9.700\,4 \times 10^2\,kg\,m^{-3}$.

MATHS HELP
SIGNIFICANT FIGURES

In calculations like the evaluation of the density of sodium above, you need to be aware of the number of significant figures of each of the quantities given, which has an implication for the accuracy of your answer. This should be given to the same number of significant figures as the quantity with the smallest number of significant figures provided in the data for the question. However, you should only 'round down' to this number of significant figures at the end of the calculation. In this example, the relative atomic mass of sodium is given to six significant figures, whereas the unit cell dimension and the Avogadro number are given to five significant figures, which is why the crystal density at the end is quoted to five significant figures.

Table 3.1 Primitive and centred lattices

Lattice name	Symbol	Number of atoms in unit cell
primitive	P	1
body-centred	I	2
face-centred	F	4

3.3 Summary of Section 3

1   A crystal is defined as a solid with a regularly repeating internal structure.

2   The internal periodicity of crystals was predicted in the seventeenth century from the shapes of crystals and the constancy of the interfacial angles for particular minerals.

3   The repeating unit in a crystal is called the unit cell.

4   The crystal lattice is a three-dimensional array of points throughout the structure, all of which have identical environments.

5   There are seven unit cell shapes which are completely space filling when packed together; the three discussed in this Book are the cubic, tetragonal and orthorhombic cells.

6   Primitive, P, body-centred, I, and face-centred, F, unit cells have different numbers of repeat units (atoms, ions or molecules) in the unit cell: 1, 2 and 4, respectively.

7   From a knowledge of the size and contents of a unit cell, the density of a crystal can be calculated.

QUESTION 3.2

A certain element, X, crystallizes in a body-centred cubic unit cell of dimension 286.6 pm. The density is $7.874 \times 10^3 \, \text{kg m}^{-3}$. What is the element?

QUESTION 3.3

Figure 3.18 shows a repeating pattern of cubes and stars, with five areas highlighted as A–E. How many of these outlines are unit cells? What type of unit cells are they?

QUESTION 3.4

The unit cell of caesium metal is shown in Figure 3.19. How many atoms of caesium are there in the unit cell?

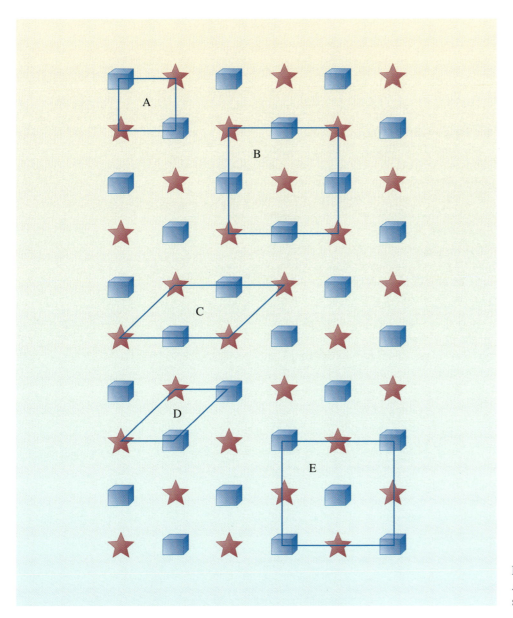

Figure 3.18
A repeating pattern of cubes and stars.

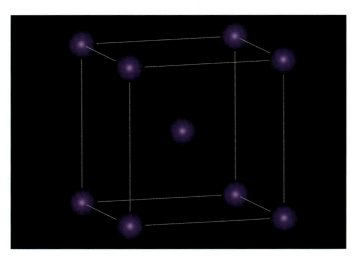

Figure 3.19 The unit cell of caesium metal.

IONIC SOLIDS

4

Elements in the Groups at the far left and far right of the Periodic Table readily form ions. Thus, we expect the metals in Groups I and II to form positively charged ions —**cations**—and the non-metals of Groups VI and VII to form negatively charged ions—**anions**—because by doing so they are able to achieve the stable electronic configuration of a noble gas. Cations can also be formed by some of the Group III elements such as aluminium, Al^{3+}, by the transition metals and even occasionally by the high atomic number elements in Group IV, such as tin and lead, which can form Sn^{2+} and Pb^{2+}, and Sn^{4+} and Pb^{4+} ions (each successive ionization of an electron becomes more difficult because the remaining electrons become more strongly bound, and highly charged ions such as these are more unusual; this is discussed further in *Elements of the p Block*[2]). In Group V, nitrogen can form an anion, N^{3-}.

An **ionic bond** is formed between two oppositely charged ions. The attraction between them is given by the balance of the charges present. The electrostatic force, F, between two oppositely charged ions is governed by **Coulomb's law** (Equation 4.1) and is proportional to the product of the two charges, q_1q_2, and inversely proportional to the square of the distance between them, r^2; that is,

$$F \propto \frac{q_1q_2}{r^2} \qquad\qquad (4.1)$$

This is known as an **inverse square law**. The force is not dependent on the mass of the ions, but only on the charges. When the distance between the oppositely charged ions is very small, the attractive force is balanced by repulsive forces, which originate from interactions between the nuclei and the electrons of the two species. Ionic bonds are strong and non-directional; their strength decreases with increasing separation of the ions, but they are still effective over long distances.

The inverse square law also applies to ions of like charge, but in this case the force between them is repulsive in character.

Ionic crystals are therefore composed of infinite arrays of ions, which pack together in such a way as to minimize the total energy of the system. This is achieved by maximizing the Coulombic attraction between oppositely charged ions and minimizing the repulsions arising from interactions between like ions, and also from the forces that generally arise when two or more chemical species* are sterically crowded. We expect to find ionic compounds in the halides and oxides of the metals in Groups I and II, where valence electrons from the metal are *transferred* to the non-metal. It is with such crystal structures that this Section begins.

However, just because it is possible to form a particular ion, it does not mean that this ion will always exist whatever the circumstances. In many structures we find that the bonding is not purely ionic but that it possesses some degree of **covalency**, where the electrons are *shared* between the two bonding atoms rather than transferred from one to the other. This is particularly true for the main-group elements in the centre of the Periodic Table. This point is taken up later: in Section 5, we discuss

---

* This word is an umbrella term embracing all chemical entities — atoms, molecules, ions and radicals.

the sizes of ions and the limitations of the concept of ions as hard spheres. In Section 6, we go on to look at some covalently bonded crystal structures.

4.1 Structures with the general formula MX

● You have met some ionic compounds with the general formula MX before: give some examples.

◐ Chlorides of the alkali metals, such as NaCl, KCl, and CsCl, fall into this category.

4.1.1 The caesium chloride (CsCl) structure

Caesium chloride, CsCl, lends its name to this particular type of crystal structure. The conventional unit cell of the **caesium chloride structure** is illustrated in Figure 4.1a. It has a caesium ion (Cs^+) in the centre of a cube with a chloride ion (Cl^-) at each corner. (Note that we could also have drawn the unit cell the other way round, with a Cl^- ion at the centre of the cube and Cs^+ ions at the corners; this would generate the same structure.)

The structure consists of two interpenetrating primitive cubic arrays, one of Cs^+ ions and one of Cl^- ions; see Figure 4.1b. (A common mistake is to describe this structure as body-centred: although all the atoms are in the same relative positions in the cube as in the *bcc* structure, the environment of the caesium at the centre of the cell is *not* the same as the environment of the chlorides at the corners. Both Cs^+ and Cl^- ions can be described as lattice points, and so either can be chosen to construct the lattice. If Cl^- is chosen, a Cs^+ ion is not a lattice point in this lattice, and therefore this is not body-centred cubic, and vice versa.)

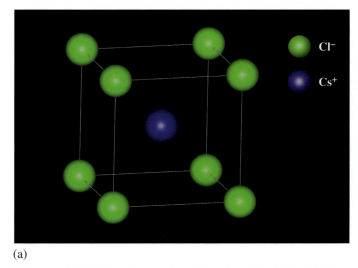

(a)

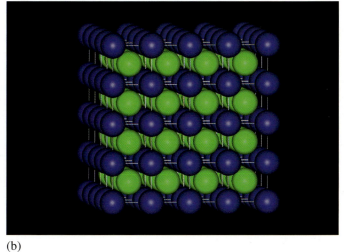

(b)

Cl^-
Cs^+

Figure 4.1 (a) The unit cell of caesium chloride, CsCl; (b) the extended CsCl structure, based on an alternative unit cell.

QUESTION 4.1

What are the coordination numbers of Cs^+ and Cl^- in Figure 4.1?

QUESTION 4.2

How many Cs^+ ions and Cl^- ions (and therefore formula units) of CsCl are there in the unit cell?

○ Cs^+ is a large cation and so is able to coordinate eight chloride ions around it. Suggest any other compounds that might crystallize in a similar fashion.

○ CsBr and CsI also have this structure. Other compounds that adopt it are NH_4Br and NH_4Cl, which contain the large ammonium ion, NH_4^+. The name 'caesium chloride' is given to this particular structure, and so they are all said to crystallize with the caesium chloride structure.

The compounds that adopt this structure tend to have appreciable ionic character and large monovalent cations. With ionic structures, individual molecules are not distinguishable because individual ions are surrounded by ions of the opposite charge.

BOX 4.1 Sal ammoniac

All ammonium compounds are very soluble in water, so surprisingly, NH_4Cl, occurs naturally as the mineral sal ammoniac (Figure 4.2). It forms on volcanic rocks near open vents and in underground burning coal seams. It was also made in ancient Egypt from the soot formed from burning camel's dung. Alexander the Great is said to have found sal ammoniac crystals in a cave in what is now Tadzhikistan — a region plagued by underground fires. The crystals are rather small with a delicate, fragile beauty: they are greatly prized by collectors because of their tendency to disappear in the first downpour!

Figure 4.2
A sample of sal ammoniac.

4.1.2 Unit cell projections — packing diagrams

It isn't always easy to represent three-dimensional structures on paper, because they can be quite complex. It is a good idea to master the art of drawing two-dimensional plans (or projections) of them, in the same way that an architect does for building plans. (In crystallographers' jargon such projections are known as **packing diagrams**, because they show you how the atoms or molecules 'pack' together to form the crystal.)

Positions of atoms or ions in crystal structures are described by **fractional coordinates**. These are coordinates based on the unit cell axes x, y, and z (known as the crystallographic axes), and are expressed as *fractions* of the unit cell dimensions a, b and c. As a system, it has the virtue that atomic positions within a unit cell can be compared from structure to structure, regardless of variation in unit cell size. As a hypothetical example, in a cubic cell with $a = 1\,000\,\text{pm}$, if an atom occurs halfway along the unit cell in the x direction at $x = 500\,\text{pm}$, then its fractional coordinate in the x direction is given by

$$\frac{x}{a} = \frac{500}{1\,000} = 0.5$$

Similarly, in the y and z directions the fractional coordinates are given by y/b and z/c.

The packing diagram for the CsCl unit cell in Figure 4.1a is shown in Figure 4.3. This represents the projection obtained when looking down the z crystallographic axis of the unit cell. This projection is on to the xy plane. Thus, the x and y coordinates are not recorded because they can simply be represented by their position in the plane of the paper. The z coordinate is recorded against an atom or ion if it is fractional; thus, the Cs^+ ion in the centre of the projection is marked at height 0.5, as its position is half way up the unit cell. An atom or ion with $z = 0$ lies on the bottom face of the cell, and must, by the repetitive qualities of the unit cell, occur again at $z = 1$, on the top face of the unit cell. The green circles at the corners of the square in Figure 4.3 represent chloride ions at $z = 0$ (and $z = 1$). Note that it is conventional to mark only fractional coordinates — that is, not 0 or 1.

4.1.3 The sodium chloride (NaCl) structure

Sodium chloride, or common salt, is probably the most familiar chemical of all: it is used by most of us on food or in cooking every day and has a very characteristic taste. Indeed the word 'salt' is used generically by chemists to describe the whole group of ionic compounds: in general, a **salt** is one product of a neutralization reaction between an acid and a base (the other product being water). Seawater contains about 3% salt (Na^+ 10 500 mg litre^{-1} and Cl^- 19 500 mg litre^{-1}). The mineral is known as halite or rock salt (Figure 4.4a), and occurs as large evaporite deposits, which are the remains of ancient dried-up salt lakes and seas (Figure 4.4d).

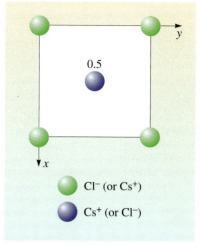

Figure 4.3
The unit cell projection of the CsCl unit cell shown in Figure 4.1a.

(a)

(b)

(c)

(d)

Figure 4.4   (a) Naturally occurring sample of halite; (b) Goat Mineral Lick on the Icefields Parkway, Alberta, Canada; (c) the Blessed Kinga Chapel carved into one of the chambers at the Wieliczka salt mines, Cracow, Poland (even the chandeliers are made of salt crystals!); (d) salt flats.

BOX 4.2 Crystals in action 2: salt

Salt is an essential mineral for life. People who eat milk products, and raw or roasted meat take in salt naturally as part of their diet, but people who eat mainly cereals and vegetables (or boiled meats) need salt supplements (the dietary requirement per day per person is 2 000–15 000 mg Na). Herbivorous animals, livestock and poultry, similarly need salt supplements and Figure 4.4b shows a mountain goat licking at a salt-rich mineral outcrop.

Because of the essential nature of salt in the diet, it has been a highly prized commodity for centuries: Roman officers received a salt allowance — the *salarium* — from which the English word 'salary' is derived.

The need for salt has even influenced the path of history, being the cause of bitter warfare. It helped to spark off the French Revolution: in eighteenth-century France, all the power and wealth was held by the nobility and the clergy, who were exempt from most taxes. The taxes on peasants were crushing and included the *gabelle* — a salt tax, which required everyone over the age of seven to buy 7 lb of salt a year. Some 10 000 peasants were imprisoned annually, and several hundred executed, for offences against the salt law alone. During the American Civil War, in 1864, the Union forces made a forced march and fought a 36-hour battle to capture Saltville, Virginia, the site of an important salt-processing plant thought essential to the armies of the South. A punitive salt tax was imposed on British-ruled India, and in the campaign for Indian independence, 'salt marches' were organized by Mahatma Gandhi to protest against the levy of this tax.

Salt is found in many countries in the world as huge underground deposits. These are the remnants of ancient seas, which have evaporated to dryness. The biggest deposits in the UK are in Cheshire, and many place names there, such as Nantwich, end in 'wich', meaning salt. Salt is obtained from underground deposits either by mining the dry mineral using conventional mining methods, or by pumping water down to dissolve the salt, extracting the brine, and then recovering the salt from the brine by various drying techniques. In hot maritime countries, salt is also made by allowing seawater to cover salt pans; it is then concentrated by evaporation using the heat from the Sun, until the salt crystallizes out and can be collected. Generations of miners in Poland have carved a national treasure in the Wieliczka salt mine in Cracow. Figure 4.4c shows one of three chapels carved into a deep salt chamber in the mine, complete with salt sculptures and bas-reliefs.

The greatest single use for salt (95%) is as a chemical feedstock for the chlor-alkali industry. Electrolysis of brine is used to produce chlorine, Cl_2, and 'caustic soda', sodium hydroxide, NaOH. In the paper industry, the sodium hydroxide is then used to break down wood fibres, and the chlorine to bleach the pulp. Chlorine is also extensively used in the manufacture of vinyl chloride and polyvinyl chloride (PVC) for the plastics industry. Salt itself is used to fix dyes in the textile industry, and to preserve and cure hides in the leather industry.

NaCl tends to crystallize as small cubes (Figure 4.5a) reflecting the cubic crystal structure (Figure 4.5b); notice that the unit cell has chloride ions at the corners and in the centres of the faces.

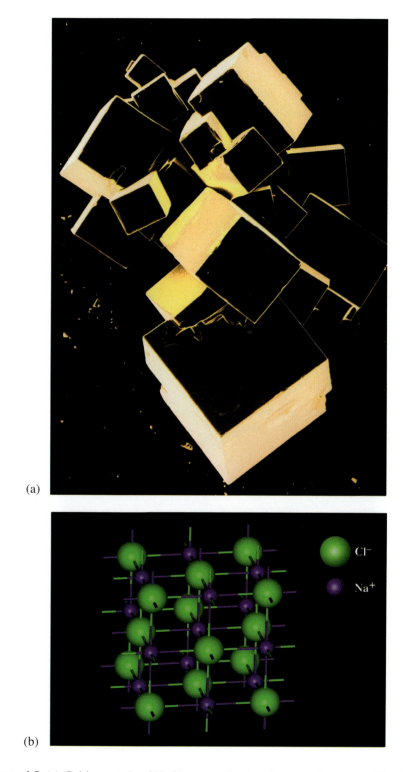

(a)

(b)

Figure 4.5 (a) Cubic crystals of NaCl seen under the electron microscope; (b) the crystal structure of sodium chloride, NaCl.

● What type of centring is this?

○ This is a face-centred, F, unit cell.

ACTIVITY 4.1 Grow some salt crystals

It is easy to grow NaCl crystals at home: simply stir salt into warm water until no more will dissolve. Leave the saturated solution on a windowsill to evaporate slowly, and small cubic crystals of NaCl will begin to form around the edges. Crystals will also tend to form on anything placed in the solution, such as a thread or twig.

Earlier in this Book you studied the solid-state structures of some metals in terms of cubic and hexagonal close-packing. In the case of simple ionic salts it is often possible to relate their structures to close-packed structures by considering one of the ions to have occupied some or all of the tetrahedral or octahedral holes within a close-packed structure.

● In a close-packed array of n atoms, how many tetrahedral and octahedral holes are there?

○ In close-packed arrays, if there are n atoms forming the array, there are $2n$ tetrahedral holes and n octahedral holes.

The **sodium chloride structure** can be described as a *ccp* array of Cl^- ions with *all the octahedral holes* filled by Na^+ ions.

● What does this description give as the ratio of Na to Cl in the structure?

○ If there are n atoms in the structure, there are n octahedral holes, and so the Na : Cl ratio will be 1 : 1. We say that the **stoichiometry** is 1 : 1.

We can now show you how the name '*cubic* close-packing' arises; the reason is that the *ccp* array has a face-centred cubic unit cell (F). Figure 4.6a shows you the relationship between the unit cell and the close-packed layers, and Figure 4.6b shows the same plane for many unit cells. In Figure 4.6c, the Na^+ ions have now been included in the octahedral holes. The close-packed layers lie perpendicular to a cube diagonal.

Note that describing the structure in this way gives us a very useful handle on the geometrical arrangements in the structure. Clearly, when an ion occupies an octahedral hole, we know that it is surrounded by six other ions in an octahedral arrangement. Later you will come across other structures, where tetrahedral holes are 'occupied'. In such cases we shall immediately know that the coordination is now fourfold and the geometry is tetrahedral. What we must not do in these descriptions is to take them literally in terms of the sizes of the ions and the shapes of the 'holes'. At this stage of the Book we have not discussed ion size, but clearly ions do vary in size, and if we looked up the actual size of a Na^+ ion, we would find that it is rather larger than the octahedral holes in a close-packed array of Cl^- ions, in which they were actually touching one another. What this type of description gives us is the *relative* geometrical arrangements of the ions in the structure.

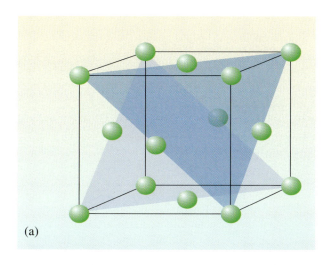

(a)

Figure 4.6 (a) A *ccp* unit cell showing the position of a close-packed layer (highlighted in blue); (b) the close-packed layer (light triangle) in a cluster of many such unit cells; (c) space-filling representation of the NaCl structure. 🖳

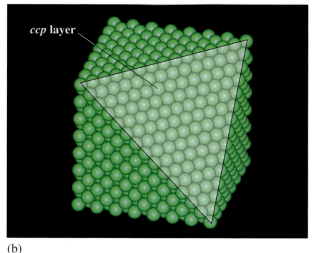

(b)

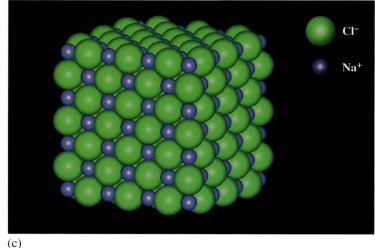

(c)

Cl⁻

Na⁺

QUESTION 4.3

Determine the coordination numbers and the geometry of the coordination around Na^+ and Cl^- in the NaCl structure.

QUESTION 4.4

Calculate the number of formula units of NaCl in the unit cell illustrated in Figure 4.5b.

Describing simple ionic structures in terms of the close-packing of one of the ions, with the other ion filling some or all of the octahedral or tetrahedral holes, can be very illuminating because it makes it easier to see the coordination geometry around a specific ion. It also enables you to see where there are spaces in the structure, which can be very important for transport processes, such as diffusion and ionic conduction.

Other compounds that crystallize with the sodium chloride structure include the silver halides, such as AgCl, all the other alkali metal halides (apart from CsCl, CsBr and CsI), and the alkaline earth metal oxides and sulfides, such as MgO.

Different ways of describing crystal structures can be useful because they can bring out a particular feature or relationship between the atoms. For sodium chloride we have seen it described in two ways: (i) using a picture of its unit cell, and (ii) using

STUDY NOTE

You can see this structure being built in the 'Building the NaCl structure' section of *Model Building* on one of the CD-ROMs associated with this Book.

The NaCl structure is also illustrated towards the end of the *Crystals* CD-ROM program.

a description in terms of close-packing. We saw in Question 4.3 that each Na^+ ion is surrounded by an octahedron of six Cl^- ions, now illustrated in Figure 4.7. A third way of interpreting the NaCl structure is by looking at how these octahedra link up. The NaCl structure can be described in terms of $NaCl_6$ octahedra sharing edges. An octahedron has twelve edges, and each one is shared by two octahedra in the NaCl structure. This is illustrated in Figure 4.8, which in (a) shows a NaCl unit cell with three $NaCl_6$ octahedra picked out in blue: a maximum of only six octahedra can meet at a point, and one of the resulting tetrahedral spaces (holes) is depicted by yellow shading. In Figure 4.8b, the NaCl structure is shown by linked edge-sharing and vertex-sharing octahedra; the positions of adjacent tetrahedral holes are picked out in yellow.

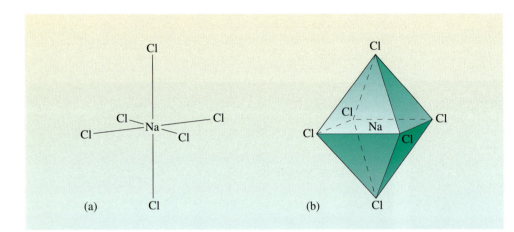

(a)

(b)

Figure 4.7 (a) An octahedron of chloride ions around a sodium ion in the NaCl crystal; (b) an $NaCl_6$ octahedron, shown as a solid shape.

Figure 4.8 (a) The NaCl structure, showing edge-sharing of (blue) octahedra (a tetrahedral space is also shown, shaded in yellow); (b) extended NaCl structure showing edge-sharing octahedra, and also depicting adjacent tetrahedral holes (in yellow).

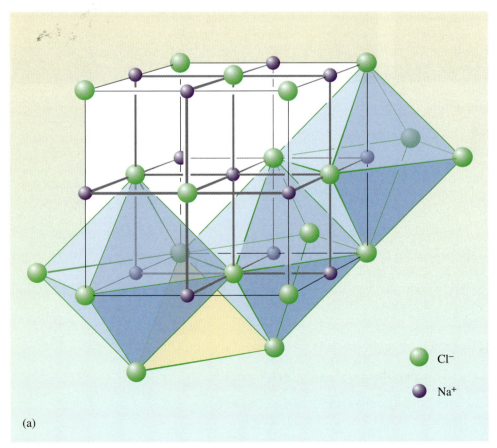

(a)

Cl⁻

Na⁺

(b)

QUESTION 4.5

Draw and label a projection of the NaCl unit cell in Figure 4.5b.

QUESTION 4.6

The NaCl unit cell in Figure 4.5b has a unit cell of dimension 564 pm. Calculate the density of crystalline NaCl in $g\,cm^{-3}$.

QUESTION 4.7

The density of AgCl was measured as $5.571\,g\,cm^{-3}$. AgCl crystallizes with the same structure as NaCl. What is the unit cell dimension?

QUESTION 4.8

AgX crystallizes with the same structure as NaCl, and is found to have a density of $6\,477\,kg\,m^{-3}$ and a unit cell dimension of 577.5 pm. Identify the element X.

QUESTION 4.9

Would a structure based on the hexagonal close-packing of anions with cations occupying the octahedral holes be possible?

4.1.4 The zinc blende (ZnS) or sphalerite structure

Zinc sulfide occurs naturally as the mineral zinc blende or sphalerite. Often it is contaminated with traces of iron; the iron gives a strong dark colour to the crystals, and it is called 'Black Jack' by the miners (Figure 4.9). The unit cell of the crystal structure is illustrated in Figure 4.10. It is based on a cubic close-packed array of sulfide ions, S^{2-}, with Zn^{2+} in some of the tetrahedral holes.

Figure 4.9 Crystals of sphalerite (Black Jack).

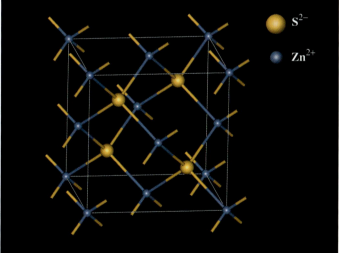

Figure 4.10 Unit cell of zinc blende.

● What proportion of the tetrahedral holes will be occupied in these structures?

● Half of them. In a close-packed array of n atoms there are $2n$ tetrahedral holes, and only n are occupied in zinc blende.

● What is the stoichiometry of this structure?

● There is one zinc for every sulfur, so the stoichiometry is 1 : 1.

The **zinc blende structure** can thus be thought of as a cubic close-packed array of sulfur ions with zinc ions occupying alternate tetrahedral holes. (Notice that if all the atoms were identical, the structure would be exactly the same as that of diamond, which we shall look at in more detail later.) Examples of compounds that crystallize in the zinc blende structure are the copper halides, CuX, and zinc, cadmium and mercury sulfides, MS.

> **STUDY NOTE**
>
> You can see this structure being built in 'The zinc blende structure' section of *Model Building* on one of the CD-ROMs associated with this Book.

QUESTION 4.10

Using either Figure 4.10, or the models in the *Model Building* program on the CD-ROM, determine the coordination numbers of Zn and S in the zinc blende structure.

QUESTION 4.11

Using Figure 4.10, draw a unit cell projection of zinc blende.

QUESTION 4.12

How many formula units, ZnS, are there in the zinc blende unit cell?

QUESTION 4.13

Would you expect there to be a hexagonal form of ZnS based on a *hcp* array of sulfide ions?

QUESTION 4.14

Determine from Figure Q.7 (p. 113) the coordination number of each type of atom.

4.2 Structures with the general formula MX$_2$

4.2.1 The fluorite and antifluorite structures

The **fluorite structure** is named after the mineral fluorspar, CaF$_2$, calcium fluoride, and is illustrated in Figure 4.11. The mineral Blue John (Figures 4.12 and 4.13) from Derbyshire has the fluorite structure. An explanation of the beautiful colours in Blue John is given in Section 8.1.1.

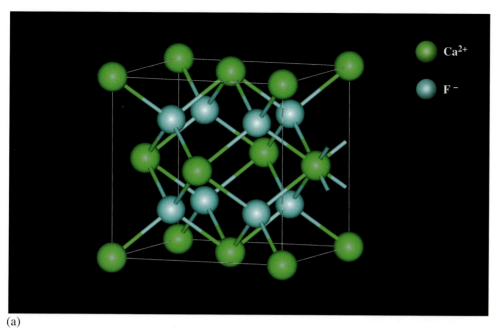

(a)

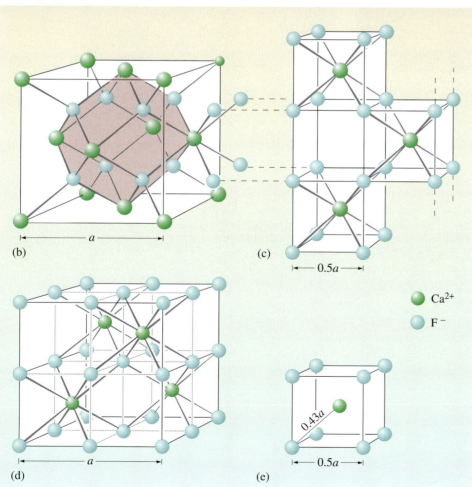

(b)

(c)

(d)

(e)

Figure 4.11
The crystal structure of fluorite,
CaF$_2$. (a) WebLab ViewerLite image
of the fluorite structure; (b) the unit
cell as a *ccp* array of cations (the
pink-shaded area indicates the
position of an octahedral hole);
(c) part of this structure redrawn
based on a primitive cubic array of
fluoride ions; (d) a unit cell based
on the arrangement in part (c);
(e) the relationship of unit cell
dimensions to the primitive anion
cube (the octant).

57

Figure 4.12 The mineral Blue John, which has the fluorite structure; the name is a corruption of the French *bleu-jaune*, which describes the main colours in the mineral.

Figure 4.13 A vase made from Blue John.

The fluorite structure can be described as a *ccp* array of Ca^{2+} ions with *all* the tetrahedral holes occupied by F^- ions.

⬤ Look at Figure 4.11, and work out the coordination numbers of the ions and the stoichiometry of the fluorite structure.

⬤ Each calcium ion is surrounded by eight fluoride ions at the corners of a cube, so the coordination number of Ca^{2+} is 8. Each fluoride ion is surrounded tetrahedrally by four calcium ions; thus, fluorite has 8 : 4 coordination. Because there is one calcium ion for every two fluorides, the stoichiometry is 1 : 2.

The cubic eight-coordination of calcium is not easy to see in Figure 4.11a and b, but it can be seen in Figure 4.11c, in which the structure has been extended beyond the unit cell and smaller cubes have been drawn with fluoride ions at each corner. In Figure 4.11d the origin of the unit cell has been moved so that this feature can be seen more clearly. For clarity, this unit cell has been divided into eight smaller cubes called octants (Figure 4.11e); alternate octants are occupied by calcium cations. Notice also that the larger octahedral holes are vacant in this structure: one of them can be seen highlighted in pink at the body-centre of the unit cell in Figure 4.11b: it is enclosed by the six Ca^{2+} ions at the centre of each face of the cube.

The **antifluorite structure** is the same structure as fluorite, but with cations replaced by anions, and vice versa. It is exhibited by lithium oxide, Li_2O, lithium telluride, Li_2Te, and the oxides and sulfides of the alkali metals, M_2O and M_2S. Fluorite and antifluorite are the only structures in which 8 : 4 coordination is found.

4.2.2 The cadmium chloride ($CdCl_2$) and cadmium iodide (CdI_2) structures

These structures are based on a *ccp* array of chloride ions and an *hcp* array of iodide ions, respectively. In both cases half the octahedral holes are occupied by Cd^{2+} ions. A layer of anions with fully occupied octahedral sites alternates with a layer containing unfilled octahedral holes. The **cadmium iodide structure** is shown edge-on to the planes in Figure 4.14a, and a bird's-eye view of the planes is shown in Figure 4.14b.

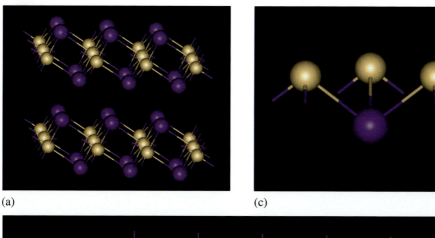

(a)

(c)

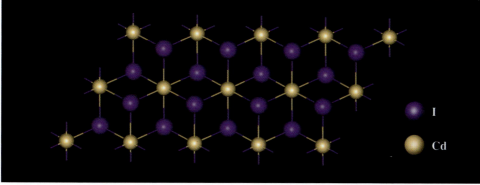

(b)

I

Cd

Figure 4.14
(a) Edge-on view of the crystal structure of cadmium iodide, CdI_2; (b) looking down on the structure of the layers in CdI_2 (also $CdCl_2$) — the halogen atoms lie in planes above and below that of the metal atoms; (c) the coordination around one iodine atom in CdI_2. 💻

Notice that Figure 4.14 shows that iodine is not completely surrounded by cadmium: each iodine has three cadmium neighbours on one side (Figure 4.14c), and three iodines in the next layer on the other. In an ionic structure, we expect each ion to be surrounded by ions of opposite charge, as in NaCl. If we regard CdI_2 as ionic, we find that I^- ions are juxtaposed with other negative iodide ions in the next layer. This suggests that we are not entirely justified in regarding the bonding in CdI_2 and similar compounds as wholly ionic. This is a point that we shall take up in Section 5.5.

QUESTION 4.15

Use Figure 4.11d to draw a packing diagram for the fluorite structure, CaF_2.

QUESTION 4.16

Look back to Figures 2.6, 2.8 and 2.10, and draw packing diagrams of unit cells for the (a) *hcp* and (b) *ccp* structures, seen perpendicular to the close-packed layers (cf. Figure 4.15a). Assuming that the close-packed layer is the *ab* plane of Figure 4.15b, draw in the *x* and *y* coordinates of the atoms in their correct positions, and mark the third coordinate, *z*, as a fraction of the corresponding repeat distance, *c*. Remember that the unit cell dimension *c* in *hcp* structures spans three layers, ABA, whereas in *ccp* it spans four layers, ABCA; hence there is one layer *between* each A layer in *hcp* and two layers between each A layer in *ccp*, so the *c* dimension in *ccp* is 50% longer than in *hcp* (Figure 4.15c).

Figure 4.15
(a) Unit cell shown marked on close-packed layer A; (b) unit cell projection for cubic and hexagonal close-packing; (c) the *c* dimension in *hcp* and *ccp*.

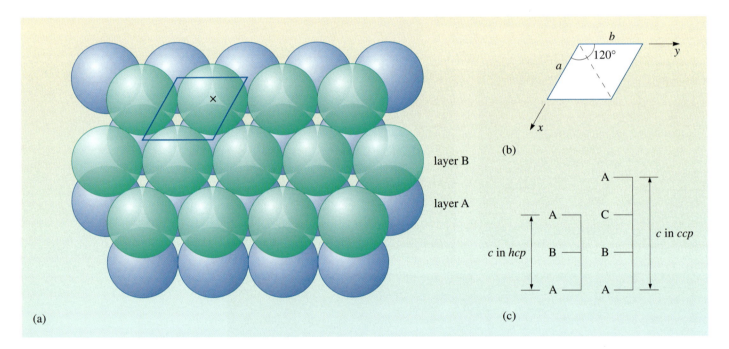

(a)

(b)

(c)

layer B

layer A

c in *hcp*

c in *ccp*

4.2.3 The rutile structure

The **rutile structure** is named after one of the mineral forms of titanium dioxide, TiO_2. A unit cell is illustrated in Figure 4.16a; the unit cell is tetragonal. This structure again demonstrates 6 : 3 coordination, but is *not* based on close-packing: each titanium atom is coordinated by six oxygens at the corners of a (slightly distorted) octahedron, and each oxygen atom is surrounded by three titaniums that lie at the corners of an (almost) equilateral triangle. It is not geometrically possible for the coordination around titanium to be a perfect octahedron, *and* for the coordination around oxygen to be a perfect equilateral triangle.

The structure can be viewed as chains of linked TiO_6 octahedra, in which each octahedron shares a pair of opposite edges, and the chains are linked by shared vertices: this is shown in Figure 4.16b.

The useful optical properties of rutile are discuused in Box 4.3 (p. 62).

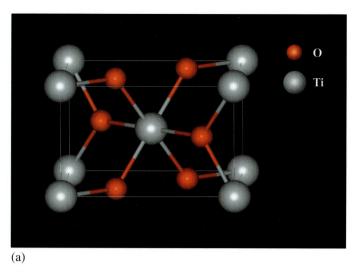
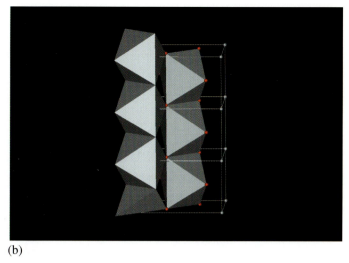

(a)　　　　　　　　　　　　　　　　　　　　(b)

Figure 4.16 The crystal structure of rutile, TiO_2: (a) a unit cell; (b) parts of two columns of TiO_6 octahedra.

QUESTION 4.17

The BiI_3 structure is based on an *hcp* array of iodine atoms, with the bismuth atoms occupying one-third of the octahedral holes. Layers with unfilled octahedral holes alternate with layers in which the octahedral holes are two-thirds occupied (see Figure 4.14 for CdX_2). What can we deduce about: (a) the coordination of bismuth; (b) the coordination of iodine?

We can generalize the calculation used in Question 4.17 to other ionic crystal structures. Take a compound with the formula A_aB_b. If the coordination number of A is C_A and that of B is C_B, then $aC_A = bC_B$. In the example of BiI_3, A is Bi, $a = 1$, B is I, and $b = 3$. We know that $C_A = 6$, because the bismuths are in the octahedral holes, so the formula gives us

$1 \times 6 = 3C_B$, and therefore $C_B = 2$

4.3 Other important crystal structures

As the valency of the metal increases, the bonding in these simple binary compounds tends to become more covalent, and the highly symmetrical structures so characteristic of simple ionic compounds occur less frequently; molecular and layer structures are common. There are many thousands of inorganic crystal structures; here we describe just another four important ones.

4.3.1 Corundum, α-Al_2O_3*

The **corundum structure** may be described as an *hcp* array of oxygen atoms with two-thirds of the octahedral holes occupied by aluminium atoms. The naturally occurring mineral is very hard (second only to diamond) and is prized for gemstones: the red variety is ruby, and other colours are known as sapphire.

---

* Aluminium oxide exists in more than one crystalline form, each designated by a Greek letter. This is also true for some other structures described in this Book; quartz and silver chloride are other examples.

BOX 4.3 Crystals in action 3: rutile

Pigments are used to give colour to coatings such as paint, plastics and so on. They are usually in the form of very small insoluble particles; the crystalline form of titanium dioxide known as rutile is the most widely used white pigment. A pigment gives a white coloration because it scatters *all* the wavelengths of visible light, rather than absorbing any. Rutile is particularly good at scattering light because it has a very high refractive index (Figure 4.17a). Figure 4.17b shows how the light is bent by rutile particles, and eventually scattered back from a painted surface layer after passing through several particles. A thick enough layer of paint with a sufficient concentration of white pigment is very good at hiding the underlying surface (Figure 4.17c). The difference between the hiding power of various white pigments is shown in Figure 4.17d.

(a)

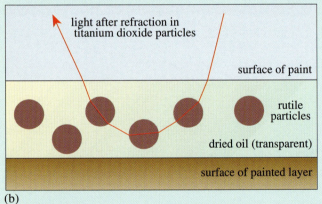

(b)

Figure 4.17 (a) Small gold-coloured needles of rutile embedded in a sample of rutilated quartz; (b) successive refractions by crystals of rutile in a painted surface; (c) white paint on a house; (d) varying effectiveness of the hiding power of some white pigments.

(c)

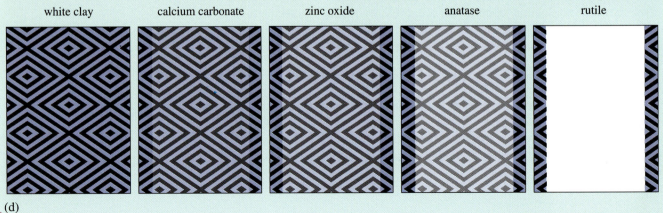

(d)

⬤ What is the coordination number of oxygen in the corundum structure?

⬤ If A = Al and B = O, then $a = 2$, $b = 3$ and $C_A = 6$. This gives:

$2 \times 6 = 3 \times C_B$, so $C_B = 4$

Once again, as we noted for rutile, geometric constraints dictate that the coordination geometry is not as simple as we might hope: if the aluminium is octahedrally coordinated, the coordination around the oxygen *cannot* be regular tetrahedral. However, it is probable that this structure is adopted in preference to other possible structures because the four aluminiums surrounding an oxygen atom approximate most closely to a regular tetrahedron. The corundum structure is also adopted by Ti_2O_3, V_2O_3, $\alpha\text{-}Cr_2O_3$, $\alpha\text{-}Fe_2O_3$, $\alpha\text{-}Ga_2O_3$ and Rh_2O_3.

4.3.2 Rhenium trioxide, ReO$_3$

The structure of rhenium trioxide, ReO_3 (also called the aluminium fluoride structure) consists of ReO_6 octahedra linked together through *each* corner to give a highly symmetrical three-dimensional network with cubic symmetry. Part of a single layer through the structure is shown in Figure 4.18a, and the linking of the octahedra in Figure 4.18b; a unit cell is illustrated in Figure 4.18c. The **rhenium trioxide structure** is adopted by the fluorides of Al, Sc, Fe, Co, Rh and Pd, and also by the oxide WO_3 (at high temperature).

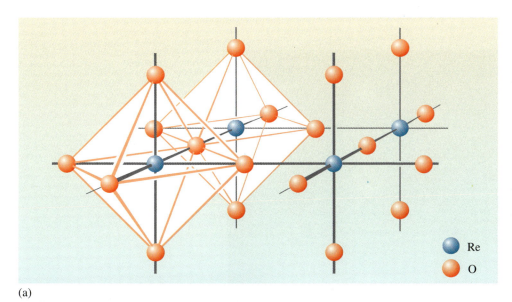

Re
O

(a)

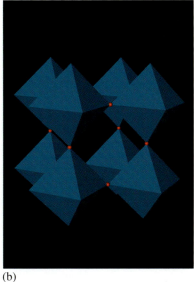

(b)

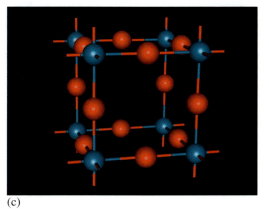

(c)

Figure 4.18 Part of the ReO_3 structure, showing linking of ReO_6 octahedra through their corners: (a) part of a single layer; (b) part of two layers, showing linked octahedra; (c) a unit cell. 🖥

4.3.3 The perovskite structure

The mineral $CaTiO_3$, which adopts the **perovskite structure**, was named after a Russian mineralogist, Count Lev Aleksevich von Perovski. A unit cell is shown in Figure 4.19: this unit cell is termed the A-type, because if we take the general formula ABX_3 for the perovskites, then in this cell the A atom is at the centre.

⬤ What is the coordination around calcium in Figure 4.19?

⬤ Calcium is coordinated to twelve oxygens (X) at the midpoints of the cell edges. There are also eight titaniums (B) at the corners of the cube.

The structure can be usefully described in other ways. It can be thought of as a *ccp* array of Ca and O atoms with the Ti atoms occupying the octahedral holes. To help you see this more easily, look at Figure 4.20, which shows a close-packed* plane (shaded blue) consisting of the central Ca atom and six surrounding O atoms, together with the planes above and below this, each containing three O atoms.

Alternatively, perovskite can be viewed as having the same octahedral framework as ReO_3, based on BX_6 (TiO_6) octahedra, but with an A atom added in at the centre of the cell. Other compounds adopting this structure include $SrTiO_3$, $SrZrO_3$, $SrHfO_3$, $SrSnO_3$ and $BaSnO_3$. The structures of a series of important compounds known as **high-temperature superconductors** are based on the perovskite structure (see Box 4.5, p. 66).

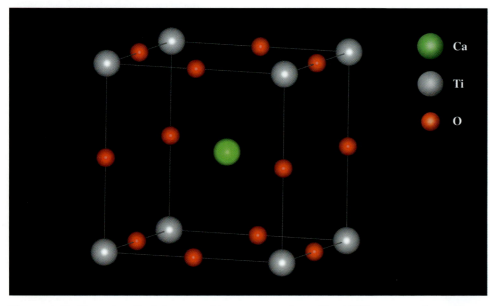

Ca

Ti

O

Figure 4.19 The unit cell of the perovskite structure adopted by ABX_3 compounds, such as $CaTiO_3$. 🖥

* Strictly speaking, it is not close-packed according to the definition that was given earlier, which was in terms of identical atoms; however, the geometrical arrangement is the same.

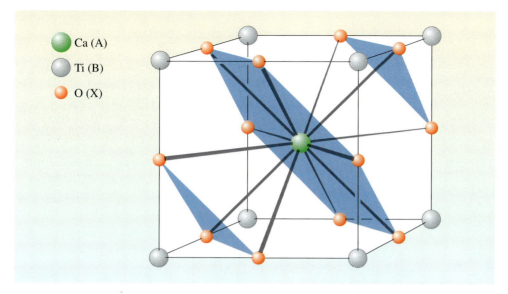

Legend:
- Ca (A)
- Ti (B)
- O (X)

Figure 4.20
The perovskite structure, with the close-packed AX planes shown in blue.

4.3.4 The spinel ($MgAl_2O_4$) structure

The **spinel structure** is rather complex; it is based on cubic-close-packing of oxide ions, with the divalent Mg^{2+} ion in tetrahedral holes and the two trivalent Al^{3+} ions in octahedral holes.

The *inverse* spinel structure has the same structural arrangement as spinel, but now one of the trivalent ions occupies the tetrahedral sites and the other M^{3+} ion and the M^{2+} ion share the octahedral sites (see also Box 4.4).

BOX 4.4 Crystals in action 4: magnetite

Magnetite, Fe_3O_4, is a naturally occurring magnetic mineral, and it can be made by the controlled partial oxidation of FeO. It contains both Fe^{2+} and Fe^{3+}, and has the inverse spinel structure.

This magnetic mineral has been known since ancient times as lodestone. Around the twelfth century, mariners in both China and Europe noticed that when a piece of lodestone (Figure 4.21a) was floated on a stick in water it tended to orientate itself to point at the Pole Star (that is, north). Once it was found that an iron needle became similarly magnetized when left in contact with the lodestone, the discovery was put to good use in early compasses, with the needle mounted on a pin in the base of the compass bowl (Figure 4.21b).

Magnetite, and related compounds known as ferrites, have recently found use in memory devices in computers, in transformer cores, and as magnetic particles on recording tapes.

(a)

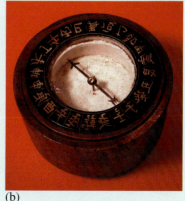
(b)

Figure 4.21 (a) Magnetite (lodestone); (b) an early compass.

BOX 4.5 Crystals in action 5: high-temperature superconductors

Understanding the perovskite structure became very important to chemists in 1986, because it was then that the first high-temperature superconductor was discovered; it was found that the structures of these unusual compounds are based on the perovskite structure.

Superconductivity was first discovered in 1911 by Kammerlingh Onnes, who noticed that when mercury was cooled to extremely low temperatures (about 4 K) using liquid helium, it lost all electrical resistance. About twenty metals and thousands of alloys were also found to exhibit this strange phenomenon, but it wasn't until 1933 that Meissner noticed that a magnet can float over these superconductors (Figure 4.22). He suggested that the superconductor expels the magnetic field, so keeping the magnet floating in space. The combination of these two properties makes superconductors potentially very useful for the transportation of electricity with no energy loss, and in the development of magnetic devices, such as levitating trains and frictionless bearings. In practice, although they were an interesting scientific phenomenon, because of the difficulty and expense of cooling to liquid helium temperatures they were mainly of theoretical interest, and of fairly specialized use, such as in the windings of superconducting magnets.

In 1986, two German chemists, Georg Bednorz and Alex Müller, discovered a mixed metal oxide of lanthanum, barium and copper, $La_{2-x}Ba_xCuO_4$, which became superconducting at 35 K, their work gaining them the Nobel Prize for Chemistry only a year later! Research in this area erupted, with many chemists all over the world involved in a race to find a *room-temperature* superconductor. The first superconductor to be prepared that became superconducting above liquid nitrogen temperatures (77 K) was $YBa_2Cu_3O_{7-x}$ (the '$-x$' in the formula means that there is some oxygen deficiency, which is critical to the superconducting properties), commonly known as YBCO.

Crystallographers who have determined the structures of these fascinating compounds have found that they all have one feature in common, that parallel planes of alternating copper and oxygen atoms spread throughout the structures. Figure 4.23 depicts the unit cell of YBCO, which can be thought of as a stack of three perovskite unit cells with some vacant oxygen positions. Figure 4.24 shows a perspective view of many unit cells, with the copper–oxygen planes shown in dark blue. Although compounds have been made that become superconducting at around 135 K, no one has yet [2001] even come close to achieving a room-temperature superconductor.

Figure 4.22 A magnet floating over a high-temperature superconductor cooled by liquid nitrogen.

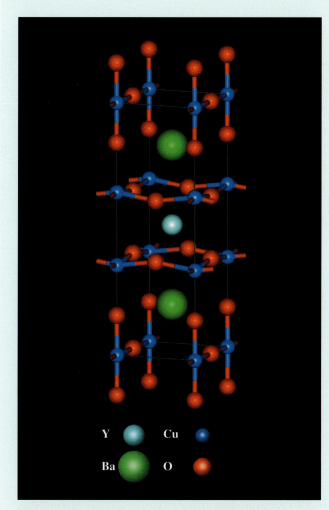

Figure 4.23 The unit cell of the high-temperature superconductor $YBa_2Cu_3O_{7-x}$ (YBCO). 💻

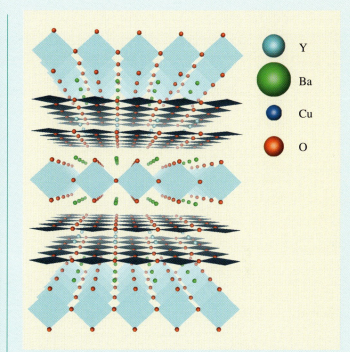

Figure 4.24 The extended structure of YBCO: the copper–oxygen planes are picked out in dark blue. 💻

The first devices using high-temperature superconductors are now on the market. These are known as SQUIDS (superconducting quantum interference devices). They are able to detect minute changes in magnetic field, and can be used by geologists to investigate such changes in the Earth's crust, and by medical researchers investigating disorders such as epilepsy (Figure 4.25).

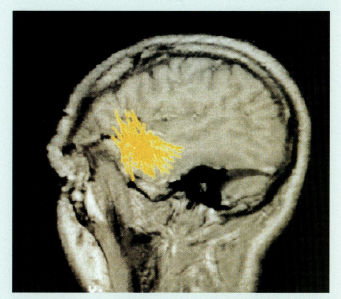

Figure 4.25 Centre of focal epilepsy as detected by a SQUID. Focal epilepsy can arise from a localized neural defect that produces magnetic signals. An array of SQUIDs measures the signals, and the inferred position of the epileptic focus (yellow) is superimposed on a magnetic resonance image.

The structures related to close-packing are summarized in Table 4.1.

Table 4.1 Structures related to close-packed arrangements

Formula	Cation : anion coordination	Type and number of holes occupied	Examples	
			Cubic close-packing	Hexagonal close-packing
MX	6 : 6	all octahedral	sodium chloride: NaCl, FeO, MnS, TiC	nickel arsenide: NiAs, FeS, NiS
	4 : 4	half tetrahedral; every alternate site occupied	zinc blende: ZnS, CuCl, γ-AgI	wurtzite: ZnS, β-AgI
MX_2	8 : 4	all tetrahedral	fluorite: CaF_2, ThO_2, ZrO_2, CeO_2	none
	6 : 3	half octahedral; alternate layers have all sites occupied	cadmium chloride: $CdCl_2$	cadmium iodide: CdI_2, TiS_2
MX_3	6 : 2	one-third octahedral; alternate layers have two-thirds of the octahedral sites occupied		bismuth iodide: BiI_3, $FeCl_3$, $TiCl_3$, VCl_3
M_2X_3	6 : 4	two-thirds octahedral		corundum: α-Al_2O_3, α-Fe_2O_3, V_2O_3, Ti_2O_3, α-Cr_2O_3
ABO_3		two-thirds octahedral		ilmenite: $FeTiO_3$
AB_2O_4		one-eighth tetrahedral and one-half octahedral	spinel: $MgAl_2O_4$; inverse spinel: $MgFe_2O_4$, Fe_3O_4	olivine: Mg_2SiO_4

4.4 Summary of Section 4

Section 4 has shown you the crystal structures of many important inorganic compounds. Wherever possible, we have tried to relate the structure to the filling of tetrahedral and octahedral holes by metal cations, in close-packed arrays of anions.

QUESTION 4.18

Draw a packing diagram for the ReO_3 structure shown in Figure 4.18c.

QUESTION 4.19

Draw a packing diagram for the perovskite unit cell shown in Figure 4.19.

QUESTION 4.20

What is the formula of a compound based on the cubic close-packing of chloride ions, with one-third of the tetrahedral holes filled by ions of the metal M?

QUESTION 4.21

A compound is based on the hexagonal close-packing of oxide ions with one-half of the octahedral holes filled by ions of the metal M. What is the valency of M in this compound?

QUESTION 4.22

What proportions of the octahedral and tetrahedral holes are occupied in the spinel structure?

IONIC RADII

5

Up to now in this Book we have regarded both atoms and ions as being hard spheres with well-defined boundaries. In close-packed arrays of atoms we assumed that they were 'touching', and we extended this idea to build up crystal structures from close-packed arrays of anions. Is this assumption justifiable? Can atoms or ions be treated as spheres of a fixed size and assigned a radius?

Ionic bonding involves the transfer of an electron from one atom to another, whereas a covalent bond is formed when electrons are shared between two atoms to form an electron pair bond; this is an idealized distinction, and in reality most bonding lies somewhere between these two extremes. Accurate crystal structure measurements from X-ray or neutron diffraction studies enable an electron density map to be plotted around each atom in a structure, showing where the electrons reside most of the time, and this can be very useful for determining the size of ions. Look at Figure 5.1, which uses contour maps to show how electron density varies between sodium and chlorine in crystalline sodium chloride (Figure 5.1a), and similarly between lithium and fluorine in lithium fluoride (Figure 5.1b). The electron density is high in the vicinity of the metal and halide nuclei, but falls off rapidly in the regions between two nuclei. Can we use this information to support a model that pictures ions as hard spheres?

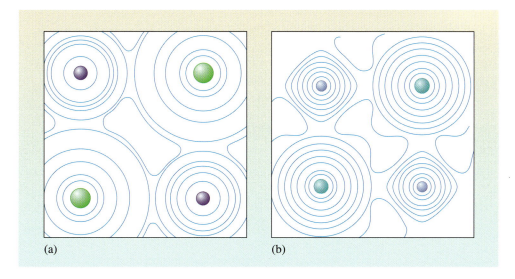

(a) (b)

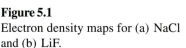

Figure 5.1
Electron density maps for (a) NaCl and (b) LiF.

The electron density distribution creates a problem. Although it falls to a small minimum value, actually there is no place on a line joining the metal and halide nuclei where it is *zero*. Thus, there is no clear boundary between the two ions: there is a finite electron density at every point on this line. If we consider the solid to be composed of ions, it looks as though the electrons of the two ions overlap to some extent, giving no well-defined boundary. This is true to a greater or lesser extent for all the simple MX metal halides.

5.1 Internuclear distances in ionic crystals

Look at the data in Table 5.1, which lists the internuclear distances between cations and the nearest anions in crystals of the sodium and potassium halides. The eight halides all have the sodium chloride structure at room temperature. If the concept of an ion as a hard sphere with a particular radius is a good approximation, we would expect the distance between the nuclei of the metal and halide ions to change by *the same amount each time* when sodium is replaced by potassium.

Table 5.1 Internuclear distances in the halides of sodium and potassium, $d(M-X)$/pm

	F^-	Cl^-	Br^-	I^-
Na^+	231	281	298	323
K^+	266	314	329	353
$d(K-X) - d(Na-X)$	35	33	31	30

● What happens to the internuclear distance when sodium is replaced by potassium?

● In all four halides, the internuclear distance increases. The increases are similar but *not* identical, varying between 35 and 30 pm.

More extensive data for the alkali metal halides are shown in Table 5.2. The coloured numbers represent the differences between the internuclear distances in adjacent compounds; if you follow any column or row, you will see again that these differ-ences are similar but not identical. For example, the differences between potassium and rubidium halides give the closest results; they are 16 pm for the fluorides, 14 pm for the chlorides, 14 pm for the bromides and 13 pm for the iodides.

Table 5.2 Internuclear distances in some alkali metal halides, $d(M-X)$/pm

	F^-		Cl^-		Br^-		I^-
Li^+	201	56	257	18	275	27	302
	30		24		23		21
Na^+	231	50	281	17	298	25	323
	35		33		31		30
K^+	266	48	314	15	329	24	353
	16		14		14		13
Rb^+	282	46	328	15	343	23	366

From the two pieces of experimental evidence that we have — the electron density plot and the changes in internuclear distances — we can see that it is not unreasonable to think of ions as being spheres with a fixed size, but we have also seen that it is not *precisely* true. We would not really expect ions to have an unvarying radius, because atoms and ions are 'squashable', and their size must be affected to a certain extent by the other ions around them. However, in spite of these reservations, it is worth while developing the idea a bit further, because the concept of an ion having a particular size — usually defined as its **ionic radius**, r — can be extremely useful.

First, we consider the relative sizes of cations and anions.

5.2 Trends in ionic radii

We shall first examine how the internuclear distances in ionic crystals vary with the charges that we assign to the ions of which they are composed. The ions Na^+, Mg^{2+}, O^{2-} and F^- *all have the same number of electrons*: they are said to be **isoelectronic**. In both NaF and MgF_2, the metal ions are surrounded by six fluoride ions at the corners of an octahedron. The internuclear distances are given in Table 5.3.

The reduction in internuclear distances from NaF to MgF_2 suggests that, although the cations have the same number of electrons, the radius of Mg^{2+} is about 30 pm smaller than that of Na^+.

When we consider anions, and compare MgF_2 with MgO, the distances suggest that the doubly charged O^{2-} ion is slightly larger than F^-.

Many examples could be cited in support of the idea that:

- the radii of isoelectronic cations decrease as the positive charge increases;
- the radii of isoelectronic anions increase slightly as the negative charge increases.

Table 5.3
Internuclear distances in NaF, MgF_2 and MgO, $d(M-X)/pm$

NaF	231
MgF_2	201
MgO	210

● How can this be related to the electronic structure of atoms and ions?

◐ We could assume that the size of ions is dependent on the spatial distribution of the electrons around the nucleus. As cations increase their charge by the loss of electrons, there is a greater positive charge to attract the remaining electrons. The repulsive forces between the electrons that are left are also eased, so these electrons occupy a smaller space around the nucleus. Conversely, acquisition of electrons by a negative ion increases the repulsive forces, which are eased by expansion of the electron distribution.

Both this explanation, and a straightforward extension of the empirical observations to include atoms, lead to the idea that *cations are considerably smaller than their parent atoms, whereas anions are larger than their parent atoms*. Thus, if we assume the existence of ions in compounds, we are led to the conclusion that, in general, cations tend to be smaller than anions for similar-sized atoms.

5.3   Determination of ionic radii

We have seen from the above discussion, that the assignment of a radius to a particular ion is not going to be a precise art. However, because the approximate size of an ion is such a useful piece of information, this uncertainty has not deterred people from making such assignments! There have been many suggestions as to how individual ionic radii can be assigned, and the chemical literature contains several different sets of values, each set being named after the person(s) who originated the method for determining the radii. We shall describe some of the methods briefly before listing the values most commonly used at present. It is most important to remember that you must not mix radii from different sets of values. Even though the values vary considerably from set to set, each set is internally consistent: if you add together two radii from one set of values you will obtain an approximately correct internuclear distance, as determined from the crystal diffraction data.

The experimental data that all systems start with are the internuclear distances determined by diffraction techniques. In order to use these distances to obtain values for individual ionic radii, the value of one radius needs to be fixed by some method. The method described in this Section is based on one of the earliest ways of doing this; it was first suggested by the German chemist, A. Landé, in 1920.

To follow the method through we shall consider those alkali metal halides that have the sodium chloride structure. The metal cations are surrounded octahedrally by six halide anions. The horizontal plane through the centre of the octahedron includes the central cation and four of the six surrounding anions, as in Figure 5.2a. Landé suggested that in the alkali metal halide with the largest anion and smallest cation, the anions must be in contact with each other, with the cation inside the octahedral hole: if the internuclear distance is known, it is then a simple matter of trigonometry to determine the anion radius (Figure 5.2b).

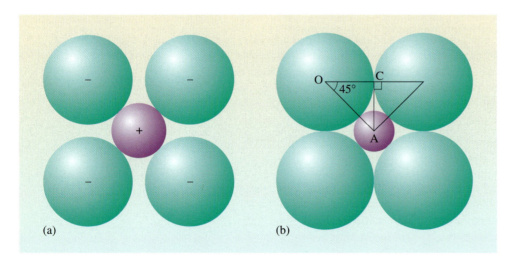

(a)   (b)

Figure 5.2 (a) Anions packed around a cation on a horizontal plane; (b) anion–anion contact on a horizontal plane through an octahedron.

The anion radius in Figure 5.2b is OC, and the angle COA = 45°. Thus

$$OC = OA \cos 45°$$

$$= \frac{OA}{\sqrt{2}}$$

where OA is the known internuclear distance.

MATHS HELP: SINE, COSINE AND TANGENT

The right-angled triangle shown in Figure 5.3 has the dimensions: angle θ, and side lengths, O (**o**pposite to θ), A (**a**djacent to θ) and H (**h**ypotenuse). We can define sine, cosine and tangent of θ in terms of these lengths:

the sine of the angle θ (written $\sin \theta$) is, $\dfrac{O}{H}$

the cosine of θ ($\cos \theta$) is $\dfrac{A}{H}$

and the tangent of θ ($\tan \theta$) is $\dfrac{O}{A}$.

These definitions can be conveniently remembered using the mnemonic, SOHCAHTOA (pronounced so-ka-toe-ah).

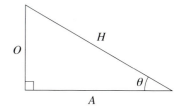

angle θ
O opposite
A adjacent
H hypotenuse

Figure 5.3
Right-angled triangle with sides O, A and H (hypotenuse).

MATHS HELP: SPECIAL TRIANGLES
ISOSCELES TRIANGLES

In the right-angled triangle shown in Figure 5.4a, the angle θ is 45°. The angles in any triangle must add up to a total of 180°, and as we know that two of the angles are equal to 45° and 90°, the remaining angle must equal 45°. This makes this particular triangle a special case of an *isosceles triangle*, in which two angles are equal, and therefore two sides must also be equal.

If we set these two equal sides to unity (1), then we can work out the length of the hypotenuse using the Pythagoras theorem (see Section 2.3, p. 26):

$$H = \sqrt{1^2 + 1^2} = \sqrt{2}$$

● What is the value of $\sin \theta$ and $\cos \theta$ for this triangle?

● For this special case, in which θ = 45°, they both take the same value,

$$\frac{1}{\sqrt{2}} = 0.707$$

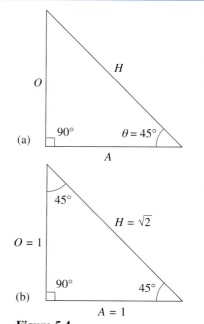

Figure 5.4
(a) A right-angled triangle with sides O, A and H (hypotenuse); (b) right-angled isosceles triangle with angle 45°.

STUDY NOTE
You can obtain further guidance on mathematical concepts from *The Sciences Good Study Guide*[1].

Although it is not possible with complete confidence to detect experimentally a situation corresponding to anion–anion contact, it is clearly most likely to occur when cations are very small, and anions are very large.

Now, the increase in internuclear distance down the columns (when the anion stays the same) of Table 5.2 suggests that cation size increases from Li^+ to Rb^+, and the increase along the rows (when the cation stays the same) suggests that anion size increases from F^- to I^-. Thus, in the context of Landé's hard-sphere model, anion–anion contact is most likely in LiI.

QUESTION 5.1

Estimate a value for the radius of the iodide ion in lithium iodide. The distance between the lithium and iodide nuclei is 302 pm.

Thus, the Landé method assigns a radius of about 214 pm for I^-; we can use this to obtain values for the radii of the other metal ions in the alkali halides, simply by subtracting it from the MI internuclear distances in Table 5.1. These values can be used in turn to obtain the radii of the other three halide anions.

There is a problem when we do this, because the variation in the difference values in Table 5.1 along a row or down a column means that the internuclear distances are *not* precisely additive; that is, we get different values for a particular ion if we use different halides.

QUESTION 5.2

Calculate a radius for F^- from the data in Table 5.2 for NaI and NaF, and the value for I^- from Question 5.1. Repeat the calculation using RbI and RbF. You should obtain two different values for the radius of the fluoride ion.

The only solution to this problem is to obtain a set of radii that gives the best statistical fit with as wide a range of experimental data as possible; this is a problem that has exercised many people since Landé!

It is very difficult to come up with a consistent set of values for the ionic radii, because, of course, the ions are *not* hard spheres: they are somewhat elastic, and the radii are affected by their environment, such as the nature of the oppositely charged ion (commonly known as the **counter ion**) and the coordination number. Linus Pauling (Figure 5.5 and Box 5.1) used a different method of calculating the radii from the internuclear distances; he produced a set of values for a hard-sphere model that is both internally consistent and which also shows the expected trends in the Periodic Table. These values are usually known as **effective ionic radii**, and they have been widely used for many years.

More recently, it has been possible to use X-ray diffraction techniques to determine accurate electron density maps (such as those in Figure 5.1) for ionic crystal structures. It was suggested that the position at which the electron density falls to a minimum should be taken as the radius for each ion because it represents most closely the actual physical size of an ion. These experimentally determined ionic radii are often called **crystal radii**: the values are somewhat different from the older sets, and tend to make the anions smaller and the cations bigger than those values.

BOX 5.1 Linus Pauling 1901–1994

Linus Pauling was born in Portland, Oregon, USA. He worked in Europe briefly before returning to an academic position at the California Institute of Technology. He used quantum mechanics to explain how atoms bond in a molecule, and his book *The Nature of the Chemical Bond*, published in 1939, is a classic text on the subject. He went on to study the chemistry of living systems, looking at the nature of hydrogen-bonding in proteins, and finally in 1948 working out the α-helix structure of a polypeptide. In 1954, he was awarded the Nobel Prize for Chemistry for his work on molecular structure, particularly the intermolecular forces in proteins.

Pauling was also a political activist. In 1953, his book *No More War* was published. During the McCarthy era he was denied a passport for several years, only finally managing to obtain one barely two weeks before his

Nobel Prize ceremony! He presented a petition to the UN containing the signatures of more than 11 000 scientists, calling for an immediate ban on nuclear testing. For these efforts he was awarded the Nobel Peace Prize in 1962 — the only person to be awarded two unshared Nobel Prizes.

Pauling's later research looked at the chemistry of the brain, the genetic causes of sickle cell anaemia, and the effects of large doses of vitamin C on the common cold and on some forms of cancer.

Figure 5.5
Linus Pauling (1901–1994).

The most comprehensive set of ionic radii has been compiled by R. D. Shannon and C. T. Prewitt. They used internuclear distances from about a thousand crystal-structure determinations to calculate their values of effective ionic radii. They used a method based on conventional values of 140 pm and 133 pm for the radii of the O^{2-} and F^- ions, respectively.

Shannon and Prewitt also gave values for the crystal radii of ions based on electron density values: these values differ by a constant factor of 14 pm from the traditional values, and are based on 126 pm for O^{2-} and 119 pm for F^-. It is generally accepted that these correspond most closely to the actual physical sizes of ions in a crystal, and so they are the values normally used when dealing with physical problems, such as the diffusion of ions through a lattice. It is these values that we list in Table 5.4; a more extensive set is given in the *Data Book* (from one of the CD-ROMs).

Table 5.4 Crystal radii of some selected ions/pm[a]

Ion	Radius	Ion	Radius	Ion	Radius	Ion	Radius
Li^+	90	Be^{2+}	59	B^{3+}	41	H^-	126
Na^+	116	Mg^{2+}	86	Al^{3+}	68	OH^-	123
K^+	152	Ca^{2+}	114	Ga^{3+}	76	F^-	119
Rb^+	166	Sr^{2+}	132	In^{3+}	94	Cl^-	167
Cs^+	181	Ba^{2+}	149	Tl^{3+}	103	Br^-	182
		Fe^{2+}	92	Fe^{3+}	79	I^-	206
Cu^+	74/91[b]	Zn^{2+}	74/88[b]	Sc^{3+}	89	O^{2-}	126
Ag^+	129	Cd^{2+}	109	Y^{3+}	104	S^{2-}	170
Au^+	151	Hg^{2+}	116	Lu^{3+}	100	Se^{2-}	184

a Values are taken from R. D. Shannon, *Acta Cryst.*, 1976, **A32**, 751, and are for octahedral coordination unless noted otherwise.

b Tetrahedral/octahedral.

However, the traditional hard-sphere model radii are still commonly used in textbooks for comparing internuclear distances. The important thing to remember is that, whichever set of values you decide to use, the ionic radii are approximately additive and will reproduce the internuclear distances with a reasonable degree of accuracy; this is what makes them useful.

Figure 5.6 shows the relative sizes of the ions of some typical elements. There are some important trends to note from Table 5.4 and Figure 5.6:

- The radii of similarly charged ions increase down a Group of the Periodic Table. Compare, for example, the values for the alkali metal cations and for the alkaline earth metal cations. The radii of the halide anions show the same trend.

- The radii of the isoelectronic cations Na^+, Mg^{2+} and Al^{3+} decrease from Na^+ to Al^{3+}. The number of electrons remains constant, but the atomic number (nuclear charge) increases; the electrons are therefore pulled in towards the positively charged nucleus and the radii decrease.

- The tendency for cations to become smaller as their charge is increased is also apparent in the case of cations of the same element (e.g. Fe^{2+} and Fe^{3+}).

- For pairs of isoelectronic anions (e.g. F^- and O^{2-}), the radius increases with increasing charge because the more highly charged negative ion has a smaller (positive) nuclear charge.

- The crystal radii increase with an increase in coordination number (e.g. Cu^+ and Zn^{2+}, Table 5.4).

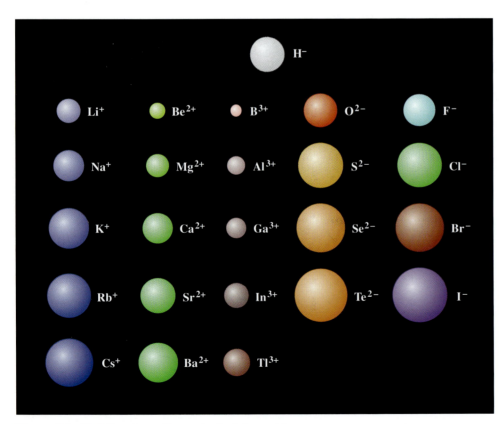

Figure 5.6 Relative sizes of some typical element ions.

5.4 Crystal structure and ionic size

In our descriptions of the ionic crystal structures in Section 4, you may have noticed variations in the coordination numbers from structure to structure.

🔵 What were those changes, and how could you account for them?

⚪ The coordination number for close-packing, where all the atoms are identical is twelve.

The coordination number of each ion in CsCl is 8.

In NaCl the octahedral holes of a *ccp* array of Cl^- ions are occupied by Na^+ ions, and the coordination number is 6 for both ions.

In ZnS, Zn^{2+} ions occupy the tetrahedral holes of a *ccp* array of S^{2-} ions, and the coordination number is 4 for both ions.

We would expect that one of the contributory factors to these changing coordination numbers is the different sizes of the cations; for example, Cs^+ ions are much larger than Zn^{2+} ions.

In an ionic compound, we argue that the Coulombic attraction between the positive and negative ions hold the structure together. At first sight, therefore, we might expect that the structure of lowest energy is the one that puts any ion into contact with the largest number of oppositely charged ions. If our reasoning is correct, then this type of structure should be the most common.

Of the MX structures that we have considered, we might expect the CsCl structure to be the most favoured one because each ion packs the largest number of oppositely charged ions around it. However, only CsCl, CsBr and CsI have the CsCl structure; the other alkali metal halides adopt a sodium chloride structure. This shows that the number of cation–anion contacts is not the only factor that affects the stability of different structures.

Suppose we have an ionic compound in which each ion is surrounded by a large number of oppositely charged ions. It is convenient to use the sodium chloride structure as an example. As we pointed out in Section 5.3, the horizontal plane of the octahedron of anions around each cation includes the central cation and four of the six surrounding anions, as in Figure 5.2a. If the size of the cation is steadily decreased, a point will be reached when the anions will touch one another (Figure 5.2b). At this stage, any further decrease in cation size will no longer decrease the distance between the centres of the surrounding anions, and therefore will not increase the attraction between the ions in the lattice. Because the contact between the cation and the anions is now broken, the lattice will not be stable and the Coulombic energy can drop only if the cation adopts a smaller coordination number. We would then predict that a structure with a *lower* coordination number, such as the zinc blende structure, must be adopted if full cation–anion contact is to be maintained.

Let us reverse the progression in coordination number and consider an ionic structure in which the cations are surrounded by a relatively low number of anions, say, four. If the cation radius, r_+, is increased, there will come a time when the **radius ratio**, r_+/r_-, is large enough for six anions to be packed around the cation without anion–anion contact occurring. The structure of higher coordination number will then be the more stable. Further increases in cation size will allow *eight* anions to be packed around the cation without any anion–anion contact, and, if our arguments are valid, a new structure will be adopted.

One important specific conclusion here is that, as the ionic radii of the alkali metal and the alkaline earth metal cations increase down their Groups, coordination numbers in the compounds formed with a particular anion should also increase down the Groups. Does this actually happen in practice? Consider the alkali metal chlorides: the lithium, sodium, potassium and rubidium compounds have the six-coordinate NaCl structure, but the caesium compound has the eight-coordinate CsCl structure. This is also true of the alkali metal bromides and iodides. So the predictions hold true here. This looks promising, so how far can we take these arguments?

Still considering the ions as hard spheres, we can use simple geometry to predict the value of the radius ratio at which we would expect these changes to happen. Look again at Figure 5.2, which shows a plane through the centre of an octahedrally coordinated metal cation. The stable situation is shown in Figure 5.2a, and the limiting case for stability, when the anions are touching, in Figure 5.2b. The anion radius, r_-, in Figure 5.2b is OC, and the cation radius, r_+, is (OA − OC). From the geometry of the right-angled triangle we can see that $\cos 45° = OC/OA = 0.707$. The radius ratio, r_+/r_-, is given by

$$\frac{OA - OC}{OC} = \frac{OA}{OC} - 1 = 1.414 - 1 = 0.414$$

Using similar calculations, it is possible to calculate limiting ratios for the other geometries; these are summarized in Table 5.5.

Table 5.5 Range of radius ratios, r_+/r_-, for different coordination numbers

Coordination number	Geometry	Radius ratio range	Possible structures
4	tetrahedral	0.225–0.414	wurtzite, zinc blende
6	octahedral	0.414–0.732	NaCl, rutile
8	cubic	> 0.732	caesium chloride, fluorite

On this basis, we would expect to be able to use the ratio of ionic radii to predict possible crystal structures for a given substance. But does it work?

Unfortunately, it only works in about 50% of cases, and so is not very useful for predicting structures. The reason for this is that many factors contribute to the stability of a crystal structure.

● What factors have we considered so far?

● The total number of anions surrounding a cation, and the size of the cation.

● What else might affect the structure adopted?

● The bonding between the atoms, from wholly ionic (non-directional) to wholly covalent (directional), and the strength of that bonding will also be important.

The model so far has taken ions to be hard spheres, but this is far from the truth, of course. The picture of ions as hard spheres works best for fluorides and oxides, both of which are small, fairly incompressible anions. As atom size increases, the valence electrons are further away from the nucleus, and they are also shielded from it by the inner core electrons, and the nucleus has less influence over them. With increasing ionic radius, ions are more easily compressed — the electron cloud is more easily distorted — and they are said to have a greater **polarizability** (Figure 5.7). The ability of an ion to distort another ion, its **polarizing power**, is higher for small ions and for ions with high ionic charge. If an anion is polarized under the influence of a cation, the distribution of electrons between them becomes more directional, and this means that the bonding involved is not truly ionic but involves some degree of covalency. The higher the notional charge on a metal ion or the smaller its radius, the greater will be the covalent character of the bonding between the metallic cation and its anions. Where there is a considerable degree of covalent bonding, we often find that four-coordinate structures are stable (for example, the zinc blende structure).

It also seems that there is little energy difference between the six-coordinate and eight-coordinate MX structures; so that when, on size considerations alone, we might expect that an eight-coordinate structure would be stable, we often find that the six-coordinate structure is preferred. Eight-coordinate structures are rarely found: for instance, there are *no* eight-coordinate MO oxides. The preference for the six-coordinate NaCl structure in ionic compounds is thought to be due to the fact that the electron orbitals that are mainly involved in the bonding (the three p orbitals at right-angles to one another) are well placed for the significant overlap of the orbitals that is necessary for bonding to take place. The potential overlap of the p orbitals in the caesium chloride structure is less favourable. (The theory of molecular orbitals is the subject of *Molecular Modelling and Bonding*[3].)

So, attempts to predict crystal structures from ionic radius ratios meet with only limited success: too many other factors are involved for this to be a foolproof method. Nevertheless, the underlying trends due to ionic size can sometimes be observed, in that we expect a large cation to be more capable of higher coordination numbers than a small one: witness Cs^+, which is eight coordinate in CsCl, compared with Na^+, which is six coordinate in NaCl.

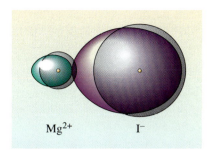

Figure 5.7
The effect of polarization on the Mg^{2+} and I^- ions (the undistorted ions are indicated in grey).

5.5 Moving away from the ionic model

Think back to Section 4.2.2, where you met the cadmium chloride and cadmium iodide structures. In these layer structures the cadmiums are surrounded octahedrally by six halogens, but the halogens have three cadmiums on one side *and none on the other*.

This type of structure is not what would be expected for an ionic solid. Describing the structure in terms of ions, the anions are *not* surrounded as far as possible with cations, but have three cations on one side and three *anions* on the other.

Let's turn our attention to some of the halides of the alkaline earth metals. The structures of barium fluoride, BaF_2, and strontium chloride, $SrCl_2$, are of the typically ionic fluorite type. Magnesium chloride, $MgCl_2$, by contrast, has a **layer structure**, which, as we have said, indicates some covalent character.

Table 5.6 shows the structures of some Group II metal halides. Those structures described as 'non-layer' are types that we shall not discuss in this Book. They are all salt-like structures, in that each ion is surrounded by ions of another type: the $CaCl_2$ and $CaBr_2$ structures are similar to rutile, with 6 : 3 coordination; the others are less regular, with the coordination of the metal being 7, 8, or 9. Those described as 'layer' have either the $CdCl_2$ or the CdI_2 structure.

Table 5.6 Structures of some alkaline earth metal halides

	F	Cl	Br	I
Mg	rutile	layer	layer	layer
Ca	fluorite	non-layer	non-layer	layer
Sr	fluorite	fluorite	non-layer	non-layer
Ba	fluorite	fluorite	non-layer	non-layer

The layer structures are concentrated towards the top right-hand corner of Table 5.6 (printed in green). For example, moving from magnesium fluoride to magnesium iodide, the structure changes from ionic to layer between the fluoride and chloride. Likewise, moving from barium iodide up to magnesium iodide, the transition to a layer structure occurs between strontium and calcium. Thus, the concentration of layer structures in the top right-hand corner is in accord with the tendency of compounds with significant covalent character to form when the electronegativity difference between metal and non-metal is small, as in the top right-hand corner of Table 5.6. By contrast, where the electronegativity difference between metal and non-metal is large, as in the bottom left-hand corner, highly ionic structures are found.

This division can be rationalized in a slightly different way. Look at the effective ionic radii of the alkaline earth metal cations and halide anions in Table 5.7 (Shannon values). Layer structures tend to occur in compounds that have small cations and large anions. Large anions are polarizable: the spherical shape of the anion is easily distorted by a concentration of positive charge, such as the Mg^{2+} cation with high polarizing power, as shown in Figure 5.7. The result is a build up of electron density in the region between the two ions; this is analogous to the partial formation of a covalent bond. So, instead of the bonding being essentially electrostatic and non-directional as in an ionic structure, it assumes the directional characteristics of a covalent bond.

Polarizability, the susceptibility of an ion to distortion of its electron distribution, is related to its ionic radius; the larger the ion the more polarizable it is. In contrast, polarizing power, the ability of an ion to distort the spherical symmetry of another ion, is inversely proportional to ionic radius.

Some semi-quantitative measure of the likelihood of partial covalency is given by the ratio $r(X^-)/r(M^{2+})$. Here you can see that we are assuming that the anion is being polarized by the cation. Naturally both ions are distorted, but the small cation very much less so than the larger anion. Let's examine the ratio $r(X^-)/r(M^{2+})$ for the compounds in Table 5.6, which showed where a transition from ionic to layer structure occurs. Values of the ratio are shown in Table 5.8, and again the boundary of the green area marks the transition from an ionic to a layer structure.

Table 5.7 Effective ionic radii for the alkaline earth metals and halide ions

Ion	$r(M^{2+})$/pm
Mg^{2+}	72
Ca^{2+}	100
Sr^{2+}	126
Ba^{2+}	142

Ion	$r(X^-)$/pm
F^-	133
Cl^-	181
Br^-	196
I^-	220

Table 5.8 Values of the ratio $r(X^-)/r(M^{2+})$ for the alkaline earth metal halides

	F	Cl	Br	I
Mg	1.85	2.51	2.72	3.06
Ca	1.33	1.81	1.96	2.20
Sr	1.06	1.44	1.56	1.75
Ba	0.94	1.27	1.38	1.55

In the layer structure side (the green-printed area) the ratio is in all cases greater than or equal to 2.2, whereas a ratio smaller than 1.96 indicates an ionic structure. The idea of polarization thus provides a rationale for the movement to the less ionic structures that occurs when potentially small cations and large anions are present[*].

A further factor that affects polarization is the ionic charge. The greater the charge the greater the polarization is likely to be. We have ignored charge, because we have been comparing cations of identical charge with anions of identical charge.

5.6 Summary of Section 5

1   The Landé method of determining ionic radii uses a hard-sphere model.

2   Shannon and Prewitt obtained a value of 220 pm for $r(I^-)$ using surveys of many iodide compounds. Other work by Shannon and Prewitt, based on electron density distributions, led to $r(I^-) = 206$ pm: the lower value is the one quoted in the *Data Book*; such radii are known as *crystal radii*. The difference between the two Shannon and Prewitt sets of values is 14 pm.

3   The ionic radii values given in Table 5.4 illustrate several trends:

   - The radii of similarly charged ions increase down a Group of the Periodic Table.

   - The radii of a series of isoelectronic cations decrease across a Period; for example, Na^+, Mg^{2+} and Al^{3+} decrease from Na^+ to Al^{3+}.

   - The tendency for cations to become smaller as their charge is increased is also apparent in the case of differently charged cations of the same element (e.g. Fe^{2+} and Fe^{3+}).

   - For pairs of isoelectronic anions (e.g. F^- and O^{2-}), the radius increases with increasing charge.

   - Crystal radii increase with an increase in coordination number.

4   The predictive power of ionic radius ratios, r_+/r_-, is limited by the fact that ions are polarizable.

5   A radius ratio, r_-/r_+, greater than or equal to 2.2 suggests the formation of layer, rather than ionic, structures.

QUESTION 5.3

What four factors help determine the stability of an ionic crystal structure?

QUESTION 5.4

What kind of compounds form layer structures?

* Notice that we have used effective ionic radii here. Because cations tend to be smaller and anions bigger in this scheme, the differences between the radius ratio values are more marked. The argument is still valid using crystal radii.

EXTENDED COVALENT STRUCTURES

6

You have now studied the structures of metals and ionic solids in some detail. A class of structures that we have not discussed yet is that of covalently bonded solids. To think of some examples of these, we can consider the elements.

- Name any non-metallic elements that are solids at room temperature.

- Some examples are: Group III, boron; Group IV, carbon, silicon, germanium; Group V, phosphorus, arsenic; Group Vl, sulfur, selenium, tellurium; Group VII, iodine.

The structures of all these elements apart from iodine (whose structure is discussed in the next Section) are extended arrays; we'll take carbon as our first example.

Diamond

Probably the most well-known form of crystalline carbon is **diamond** (Figure 6.1). The diamond structure is cubic, and is built on a face-centred cubic lattice (Figure 6.2).

Figure 6.1
A diamond.

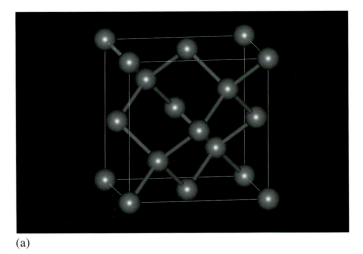

(a)

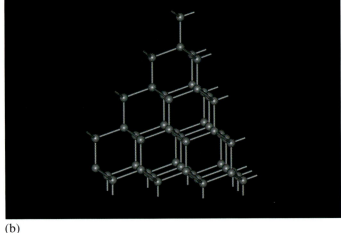

(b)

Figure 6.2 (a) A unit cell of the diamond structure; (b) alternative view of the diamond structure.

COMPUTER ACTIVITY 6.1
Investigation of the diamond structure

View the diamond structure in WebLab ViewerLite, and determine some carbon–carbon distances. Are they all the same? What is the $\angle C-C-C$ bond angle?

The Activity forms part of *Exploring the third dimension* on one of the CD-ROMs, and should take you approximately 10 minutes to complete.

All the carbon atoms in diamond are equivalent, each one forming four covalent bonds with adjacent carbons, and so are surrounded tetrahedrally by four others. The carbon atoms form an extended array, and a diamond crystal could be described as a 'giant molecule' because there are no breaks in the bonding where one molecule ends and another begins. It is interesting to note how the different type of bonding has affected the coordination: here we have identical atoms (all the same size), but the coordination number is now restricted to four because this is the maximum number of covalent bonds that carbon can form. In the case of a metallic element such as magnesium forming a crystal, where we also have identical atoms, the structure is close-packed with each atom surrounded by *twelve* others, and it is held together by metallic bonding.

⬤ Does the unit cell of the diamond structure remind you of any ionic crystal structure?

⬤ The diamond structure is geometrically the same as that of zinc blende (p. 55), but with both the zinc and sulfur atoms replaced by carbon atoms.

The covalent bonds in diamond are short and strong. Diamond is the hardest substance known and has a high m.t. (3 500 °C).

Quartz

Another familiar crystalline substance that has an extended covalent structure is silica, SiO_2. Quartz is a crystalline form of SiO_2, and is stable at room temperature. It is most commonly encountered in nature as sand, but also occurs in gemstones (Figure 6.3). If we look at the crystal structure of quartz, we do not find individual SiO_2 molecules (Figure 1.3a).

Figure 6.3
Quartz gemstones.

BOX 6.1 Crystals in action 6: synthetic diamonds

Colourless, flawless diamonds are highly prized as gemstones: they are extremely hard, fairly unreactive, and their very high refractive index (2.4) makes them disperse light, so that they sparkle brilliantly. For every gem produced, 250 tonnes of rock have to be mined and processed!

Most of the diamonds that are mined are not of gem quality as they are flawed, coloured, or too small. However, these diamonds are of vital importance to industry because of their hardness. Such diamonds are used for rock drilling bits, diamond saws, and when crushed, as the grit for grinding wheels.

Because of the rarity and expense of natural diamonds, there has always been great interest in producing synthetic diamonds. The final stimulus came with the Second World War, when diamond-tipped tools were urgently needed for cutting and machining military hardware, and it was feared that the supply from South Africa might dry up. GEC instigated a research programme, and they started to try to make diamonds from graphite (another allotrope of carbon) at very high temperatures and pressures, mimicking the conditions 160 km underground where diamonds were formed up to 3 billion years ago. It was not until they noticed the presence of diamonds in a meteorite that they came up with the idea of adding one of the minerals found in the meteorite (troilite, FeS) to dissolve the graphite, that they achieved their first success in 1951. Today, with improved techniques, almost 90% of diamonds used in industry are synthetic (Figure 6.4).

However, the hunt for synthetic gem-quality stones still goes on. In Russia, Boris Feigelson started to experiment, and by 1995 was able to grow large gems from a tiny seed diamond. These stones were flawed, as they contained small inclusions of metal, and were yellow in colour due to trapped individual atoms of nitrogen. Feigelson finally managed to refine the heating regime, so that the gems grew metal-free — but they were still yellow!

In natural diamonds, nitrogen is also trapped, but in small clusters rather than as individual atoms, such that there is no yellow coloration. Experiments on heating the diamonds under pressure to allow the nitrogen atoms to diffuse and form clusters have managed to improve the colour but not to eliminate it completely. More recently, Feigelson developed a new process that includes a nitrogen 'scavenger' — aluminium — in the mix, enabling him to manufacture large colourless gem-quality diamonds. It currently [2001] takes about 24 hours to produce a 1 carat diamond.

De Beers, the large South African diamond company, desperate to stop their trade from being devalued, have now developed a test that can distinguish these synthetic gems from the natural stones. When irradiated with ultraviolet light, only the synthetic stones phosphoresce for a few seconds after the lamp is turned off: they glow in the dark! This has been shown to be due to the way the crystals grow: the natural diamonds only show octahedral growth, but the synthetic stones contain both octahedral and cubic sectors. It is undoubtedly only a matter of time before this distinction is also overcome, and then no one will be able to tell the difference!

Figure 6.4
Mixture of natural and synthetic diamonds.

BOX 6.2 Crystals in action 7: silicon carbide

If alternate carbon atoms in the diamond structure are replaced by silicon, the **silicon carbide structure**, SiC, is obtained. SiC was discovered by Edward Acheson in 1891, when he was trying to make artificial diamonds. He had taken a mixture of clay and coke, and heated it electrically using the iron container as one electrode and carbon as the other. He first thought that the greenish crystals he found at the carbon electrode were a new compound of carbon and alumina from the clay, and named it carborundum (as distinct from 'corundum', an alumina mineral, α-Al_2O_3; see Section 4.3.1). The new

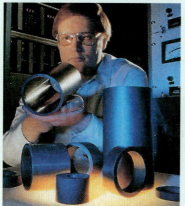

compound, which was found to be almost as hard as diamond, with very good fracture characteristics, became important industrially for grinding and polishing.

SiC can now be made by a number of advanced processes to form ceramics for special applications which need hardness, strength, low thermal expansion and chemical resistance, such as valves, bearings, nozzles and so on (Figure 6.5).

Figure 6.5 Silicon carbide construction materials.

COMPUTER ACTIVITY 6.2
Investigation of the quartz structure

Use WebLab ViewerLite to view the quartz structure and determine the Si–O distance. What is the $\angle O-Si-O$ bond angle? Are they all the same?

The Activity (in *Exploring the third dimension*) should take you approximately 15 minutes to complete.

Figure 6.6 shows part of the β-quartz structure. Notice how, once again, the covalency of each atom dictates the coordination around itself; silicon forms four covalent bonds and oxygen two.

• How can we account for the empirical formula SiO_2 for this structure?

• If every oxygen atom is attached to two silicon atoms, a silicon atom has a half share in each oxygen. As each silicon is bonded to four oxygens, then it has

$4 \times \frac{1}{2} = 2$ oxygens, and the overall formula SiO_2 results.

Quartz has a special property known as *optical activity*. This is a property of many organic molecules, and will be discussed in some detail in Part 2 of this Book, so we will not dwell on it here. Suffice it to say that this optical activity has an effect on light passing through the crystal, and arises because the crystals contain *helices* or *spirals* of linked SiO_4 tetrahedra, and these can be either all left-handed or all right-

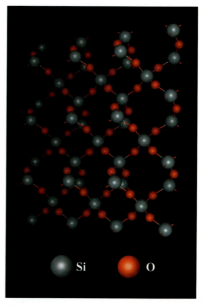

Si ⬤   O ⬤

Figure 6.6
Crystal structure of β-quartz. 🖳

handed (Figure 6.7). If you had never noticed before that a spiral has a 'handedness' then have a look at a wood screw or a corkscrew, and see which way it turns: they are both usually 'right-handed', whereas a spiral staircase usually (but not always) turns the other way. Some plants that climb by spiralling around a support can also show a particular handedness, and older students may remember a classic song from Michael Flanders and Donald Swann, which contains the lines:

> Said the right-handed Honeysuckle to the left-handed Bindweed
> 'oh let us get married if our parents don't mind we'd
> be loving and inseparable, inextricably entwined we'd
> live happily ever after' said the Honeysuckle to the Bindweed.*

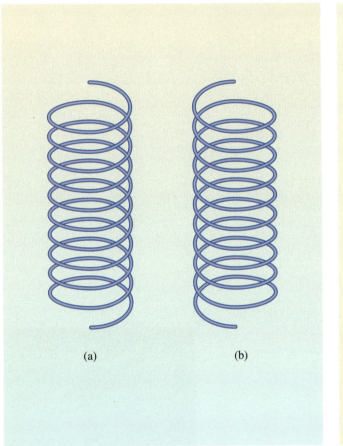

(a)   (b)   (c)   (d)

Figure 6.7 (a) Left- and (b) right-handed spirals; (c) a corkscrew; (d) a spiral staircase.

STUDY NOTE

The structures of quartz, diamond and graphite can be seen in the *Virtual Crystals* program on one of the CD-ROMs associated with this Book. The spirals of SiO_4 tetrahedra in quartz can be seen very clearly.

---

* From 'Misalliance' on the album *At the Drop of a Hat* by Michael Flanders and Donald Swann, EMI Records (1961).

BOX 6.3 Crystals in action 8: quartz

Quartz crystals possess another unusual property: if an electric field is applied across certain crystal faces, the crystal changes size; conversely, if pressure is applied to the crystal then a small voltage is produced across these faces. This is known as the *piezoelectric effect*. Because of this property, quartz crystals can be used as resonators in electronic oscillators. If a small quartz crystal plate is set up with electrodes across the faces, an applied voltage will set the crystal into a mechanical resonance of a very high frequency, thus producing an AC voltage of the same frequency. Such an oscillator is used in electronic watches and clocks (Figure 6.8). Circuits in the watch count the output signals from the oscillator to determine the passage of time, and positive feedback provides energy to the quartz crystal to keep it vibrating. The resonance frequency of well-cut crystals is very stable and well defined. Accuracy in such time-pieces is commonly one part in 10 million, but it can be much better than this under carefully controlled conditions.

Figure 6.8 A quartz watch.

Graphite

The different structural forms of an element are known as *allotropes* or **polymorphs**. Rather a different sort of crystal structure is illustrated by graphite, another allotrope of carbon. The structure illustrated in Figure 6.9 is of normal graphite (there are other graphite structures which are more complex).

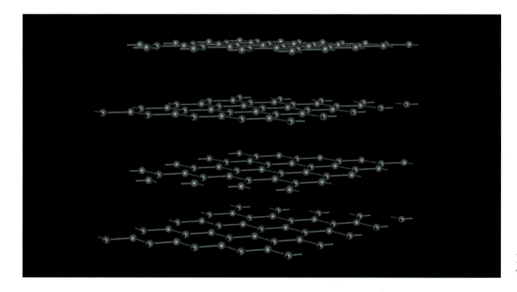

Figure 6.9
The crystal structure of graphite. 🖥

COMPUTER ACTIVITY 6.3
Investigation of the graphite structure

Use WebLab ViewerLite to view the graphite structure and determine the carbon–carbon distances, both within the layers and also between the layers. Are they the same? What is the $\angle$C—C—C bond angle in the layers?

The Activity (in *Exploring the third dimension* on one of the CD-ROMs) should take you approximately 10 minutes to complete.

You can see that the graphite structure has a giant or extended array in *two* dimensions only: there are flat sheets of carbon atoms, with each carbon atom bonded to three others at distances of 142 pm, forming a network of hexagons. The layers do not lie directly over one another, and the carbon–carbon distance *between* the layers is considerably longer, at 340 pm.

Short interatomic distances are a sign of strong bonding between atoms; conversely, long interatomic distances are indicative of weak bonds. Here we note that only weak bonding exists *between* the layers; in Section 7 we shall discuss in more detail the different types of weak bonding that hold crystals together. Graphite is a soft, grey solid, melting at the very high temperature of 3 652 °C. Its softness is attributed to the weak sheet-to-sheet bonding.

You may have noticed that the carbon–carbon distances are different in diamond and graphite.

● In which material do we find the stronger carbon–carbon bonds?

○ The carbon–carbon distances are 154 pm in diamond and 142 pm within the layers in graphite. Because the carbon–carbon distance is shorter in graphite, we are able to say that the bonding *within* the layers in graphite is stronger than that in diamond.

Graphite can be used as a lubricant in ball-bearings. The lubricant properties of graphite are due to a rather unusual reason: the small flat crystals have a layer of adsorbed nitrogen on the surface, giving a cushion that allows the crystals to glide over each other. If you put graphite in a vacuum, this gas layer is pumped off and the crystals become more sticky: graphite is not a useful lubricant in outer space!

Recently, other polymorphs of carbon, the fullerenes, have been discovered, and these are described elsewhere [2].

6.1 Summary of Section 6

1 Diamond has a face-centred cubic structure in which each C atom is covalently bonded in a tetrahedral arrangement to four other carbons.

2 Quartz, the crystalline form of SiO_2, consists of covalently bonded, linked SiO_4 tetrahedra.

3 Carbon in graphite forms sheets of carbon atoms strongly bonded in a hexagonal array, with weak bonding between the layers.

QUESTION 6.1

The carbon–carbon distances in diamond, ethene, C_2H_4, and ethyne, C_2H_2, are 154 pm, 134 pm and 120 pm, respectively. These bonds are usually classified as single, double and triple, respectively. What conclusions can you draw about the bonding in graphite?

MOLECULAR CRYSTALS

7

In Section 2, we met close-packing and looked at the structures adopted by many metallic elements. The structures that we described in Section 4 and Section 6 could be loosely categorized as 'ionic' and 'covalent', respectively. In Section 4, we found extended lattices of *ions*, in which no individual molecules can be distinguished, and for which high coordination numbers are common. In Section 6, the structures were also of the extended type (in other words the structure is continuous), but with highly directional bonding and low coordination numbers; again, it was not possible to define an individual molecule. However, as we saw in Section 5, the bonding in many crystals is not clear cut and lies somewhere between ionic and covalent. In this Section we are going to look at crystals composed of regular arrays of individual small *molecules*, each held together by strong covalent bonds. These crystals are held together by weak bonding forces, and we start this Section with an overview of the types of bonding that we find in crystals.

7.1 Bonding in crystals

First let's recap what we know about different types of bonding.

Metallic bonding

In Section 2 we looked at the structures adopted by many metals. We can think of a metal as a regular network of metal cations sitting in a sea of electrons. The electrons are free to move, accounting for the conductivity of metals, but at the same time they occupy the space between the positive ions, thus binding them together.

Ionic bonding

An ionic bond is formed between two oppositely charged ions because of the electrostatic attraction between them. The attractive electrostatic force, F, between two oppositely charged ions is governed by Coulomb's law, $F \propto q_1 q_2 / r^2$, where q_1 and q_2 are the charges on the two ions, and r is the distance between them. The force is not dependent on the mass of the ions, only on the charges.

Ionic bonds are strong and non-directional; the *energy* (force × distance) of the interaction is inversely proportional to the separation, r (the energy halves as the distance doubles). This decrease in the energy of interaction is relatively slow compared with other bonding interactions, and makes ion–ion interactions effective over large distances and the most far-reaching of the non-covalent interactions. A similar *repulsive* force is experienced by two ions of the *same* charge. Ionic crystals are composed of infinite arrays of ions, which have packed together in such a way as to minimize the total energy of the system. This is achieved by maximizing the Coulombic attraction between oppositely charged ions and minimizing the repulsive interactions. Ionic crystals are hard, brittle and non-conducting. They have high melting temperatures; in the melt they become conducting because the ions are free to move. NaCl is the archetypal example of an ionic crystal.

Covalent bonding

In many structures the bonding is covalent, the electrons are *shared* between the two bonding atoms, and there is a build-up of electron density between the two atoms. **Covalent bonds** are strong and directional. Extended covalent structures are typically hard with high melting temperatures; diamond is a prime example.

Charge–dipole and dipole–dipole interactions

Other bonding forces exist in crystals, and some of these are important for binding together small covalently bound molecules into crystalline structures.

Some elements (notably nitrogen, oxygen and the halogens) are more electronegative than others; they attract more than their share of the bonding pair of electrons from the covalent bonds they form with other elements.

⬤ How does electronegativity vary across the Periodic Table?

⬤ The electronegativity increases across a Period, and usually increases up a Group. The most electronegative element is fluorine in the top right-hand corner of the Periodic Table.

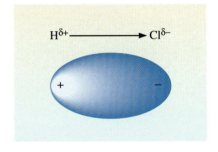

Small covalently bound molecules containing more than one type of atom can thus have an uneven distribution of electronic charge (unless they are very symmetrical in shape), such that one end of the molecule will have a slight negative charge, δ–, and the other end a slight positive charge, δ+. This separation of positive and negative charge is called an **electric dipole**, and means that the molecule can align itself in an electric field; such molecules are said to be **polar**.

⬤ In the HCl molecule, which end of the molecule will be negatively charged?

⬤ Chlorine lies just below fluorine in the top right-hand corner of the Periodic Table and so will be more electronegative than hydrogen (Figure 7.1).

Figure 7.1
The direction of the dipole in a molecule of HCl.

Molecules that possess a dipole can interact with one another (**dipole–dipole interaction**) and with ions (**charge–dipole interaction**) through the attraction of the electric charges (Figure 7.2).

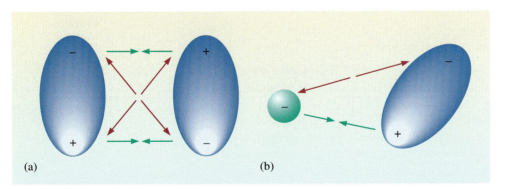

attraction
repulsion

Figure 7.2 (a) Dipole–dipole interaction: opposite ends of the dipoles attract one another; (b) charge–dipole interaction: the charge on the ion is attracted to the opposite charge on the dipole and repelled by the like charge.

The charge–dipole interaction is about 10 to 20 times weaker than an ion–ion interaction; the energy of the interaction decreases as $1/r^2$.

The energy of the dipole–dipole interaction is even weaker — about 100 times weaker than ion–ion interactions — and decreases with $1/r^3$; it thus falls off very sharply with distance.

🔵 If the distance between two dipoles doubles, by how much does the energy of interaction decrease?

⚪ $1/r^3$ becomes $1/(2r)^3 = 1/8r^3$ when the distance doubles. The energy of interaction thus falls to one-eighth of its original value.

London dispersion forces

Extremely weak forces can occur between molecules that do not possess a permanent dipole. This is because 'transient dipoles' can form in molecules due to the movement of their electrons; these dipoles can in turn *induce* dipoles in adjacent molecules. When a molecule moves near to a centre of charge or near to a dipole, an additional small dipole is temporarily set up inside the molecule, and these transient dipoles are the origin of so-called *dispersion forces*: the transient dipole in one molecule can then be attracted by the transient dipole in an adjacent molecule. The net result is a very weak, short-range attractive force between molecules known as the **London dispersion force**, after the German physicist F. London. The energy of this interaction falls off very quickly with distance, dropping as $1/r^6$.

🔵 What kind of molecules will be most likely to form London dispersion forces?

⚪ The more loosely bound electron clouds in large polarizable atoms are more likely to be susceptible to the formation of induced dipoles. Homonuclear molecules such as bromine, Br_2, and iodine, I_2, would therefore fall into this category.

Hydrogen-bonding

Although polar interactions are normally very weak, there is one special case where the interaction is strong enough for it to be exceptionally important. This is the case in which a hydrogen atom is covalently bonded to a very electronegative element such as oxygen or fluorine. The electronegative element pulls electrons towards itself to gain a slight negative charge, $\delta-$, and the H atom thus gains an equal and opposite partial positive charge $\delta+$. The positively charged $H^{\delta+}$ can now interact with the partial negative charge on the electronegative atom in an adjacent molecule, forming a weak bond and pulling the three atoms into almost a straight line. A network of alternating strong and weak bonds is built up, as in the structure of water (Figure 7.3); the weaker bonds are usually depicted by broken coloured lines, and are known as hydrogen bonds. Hydrogen bonds are particularly important in crystals and molecules in biological systems — indeed, in anything that contains water.

Table 7.1 gives some typical values for the energy of the different types of bonding discussed above.

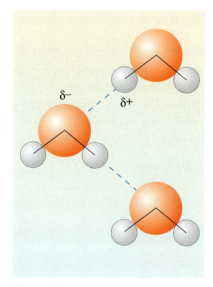

Figure 7.3
Hydrogen-bonding between water molecules.

Table 7.1 Typical values for the energy of different bonds

Type of bond	Typical energy value of bond	Example
metallic	50 kJ mol^{-1}	sodium metal, Na
ionic	500 kJ mol^{-1}	sodium chloride, Na$^+$Cl$^-$
covalent	200 kJ mol^{-1}	chlorine, Cl$_2$
charge–dipole	20 kJ mol^{-1}	salt solution, Na$^+$Cl$^-$/H$_2$O
dipole–dipole	2 kJ mol^{-1}	hydrogen chloride, HCl
London dispersion forces	< 2 kJ mol^{-1}	argon, Ar
hydrogen bond	20 kJ mol^{-1}	hydrogen fluoride, HF

7.2 Radii of atoms and ions

In Section 5, we discussed at some length the meaning of the term 'ionic radius' and how tables of such radii have been determined.

It can also be useful to know a value for the radius of an atom when it is bonded covalently, and also when it is not bonded at all. Such values can be useful when determining the size and shape of individual molecules. The size and shape of a molecule can be used in molecular modelling to help determine whether the molecule will fit into a cavity or diffuse through a channel. Such modelling is useful for drug design and in catalytic studies.

Covalent radii

The **covalent radius**, r_c, of an atom is defined as half the distance between the nuclei of two like atoms which are singly bonded, for instance, C—C in ethane, H$_3$C—CH$_3$, or chlorine, Cl—Cl (Figure 7.4a). When a suitable homonuclear diatomic molecule of the atom in question, say chlorine, exists, the determination of the radius is unambiguous. For carbon, the C$_2$ molecule does not exist except under very unusual circumstances, and so other molecules that contain the C—C bond, such as ethane are used. As the bond lengths may vary slightly from compound to compound, the accepted value of r_c is an average value taken from a number of compounds.

van der Waals radii

Dipole–dipole interactions, whether between permanent dipoles or arising from induced dipoles, are usually thought of as *non-bonded interactions*, and are usually referred to collectively as **van der Waals forces**, after the Dutch chemist Johannes van der Waals. The **van der Waals radius** of an atom, r_W, is defined as being 'a non-bonded distance of closest approach'. van der Waals radii are measured as half the shortest distance between the nuclei of two non-bonded like atoms in a crystal structure (Figure 7.4b). Such measurements are not precise, as a decision has to be taken as to whether a bond exists or not, or whether such atoms are as close together as possible. van der Waals radii are average values compiled from large sets of crystal data.

Table 7.2 shows covalent and van der Waals radii for the typical elements; the *Data Book* (from one of the CD-ROMs associated with this Book) contains the same data.

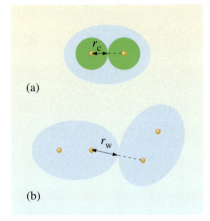

(a)

(b)

Figure 7.4
(a) The covalent radius and
(b) the van der Waals radius for
a homonuclear diatomic molecule.

Table 7.2 Single-bond covalent and van der Waals radii (in parentheses) for the typical elements in picometres

Group I	Group II	Group III	Group IV	Group V	Group VI	Group VII	Group VIII
			H 37 (120)				He — (140)
Li 135	Be 90	B 80	C 77 (170)	N 74 (155)	O 73 (152)	F 71 (147)	Ne — (154)
Na 154	Mg 130	Al 125	Si 117 (210)	P 110 (180)	S 104 (180)	Cl 99 (175)	Ar — (188)
K 200	Ca 174	Ga 126	Ge 122	As 121 (185)	Se 117 (190)	Br 114 (185)	Kr — (202)
Rb 211	Sr 192	In 141	Sn 137	Sb 141	Te 137 (206)	I 133 (198)	Xe — (216)
Cs 225	Ba 198	Tl 171	Pb 175	Bi 170	Po 140	At —	Rn —

7.3 Molecular crystals

Crystals that are composed of individual molecules held together solely by van der Waals forces are known as **molecular crystals**. It takes only a small amount of energy in the form of heat to overcome these weak intermolecular forces, and so molecular crystals are characterized by low melting temperatures. Most organic compounds form crystals of this sort, as do many inorganic compounds.

Iodine and chlorine

The simplest of molecules are diatomic molecules, which form crystals if they are cooled sufficiently. The molecules often close-pack, but they rotate freely within the crystal, behaving like spherical atoms. Hydrogen, H_2, nitrogen, N_2, and carbon monoxide, CO, all have an *hcp* structure, whereas oxygen, O_2, and fluorine, F_2, adopt a *ccp* structure. Chlorine, bromine and iodine, however, adopt an ortho-rhombic structure (see Figure 3.13a) containing non-rotating diatomic molecules. Iodine is a dark purple crystalline solid at room temperature (Figure 7.5 a and b), whereas bromine is a brown corrosive liquid and chlorine is a greenish-yellow gas (Figure 1.1b). If chlorine and bromine are cooled sufficiently, they also form crystalline solids; all three solids have the same crystal structure (although, of course, the unit cell size and the halogen–halogen distance are different for each substance). Figure 7.5c and d show the unit cell and packing diagram for the chlorine structure.

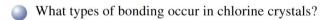 What types of bonding occur in chlorine crystals?

The structure consists of discrete diatomic molecules Cl—Cl, with strong covalent bonding between the atoms. The molecules are held together in the solid by very much weaker London dispersion forces, usually referred to as van der Waals forces.

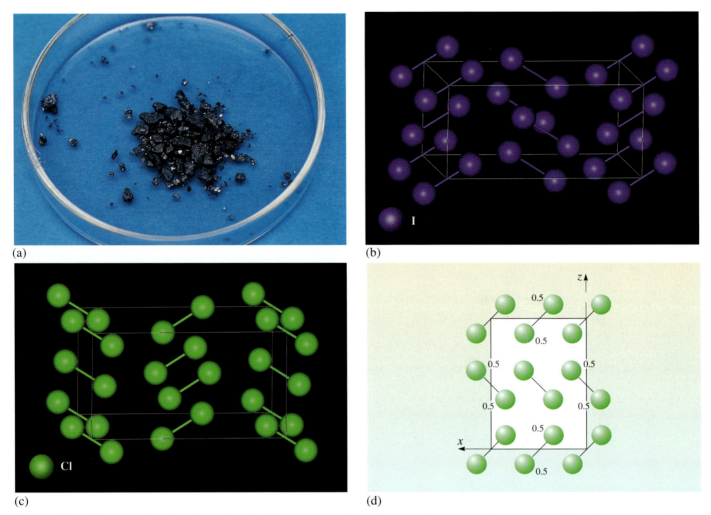

(a)

(b)

(c)

(d)

Figure 7.5 (a) Crystals of iodine; (b) unit cell of the orthorhombic iodine crystal structure; (c) unit cell of the orthorhombic chlorine crystal structure; (d) packing diagram for the orthorhombic chlorine crystal structure projected on a plane perpendicular to the y axis; atomic positions are shown, and the heights are indicated in units of the fractional coordinate b (remember that by convention we only show fractional positions). Note that in part (c) the axes are rotated anticlockwise by 90° relative to those in part (d). 💻

COMPUTER ACTIVITY 7.1 Chlorine and iodine

Use WebLab ViewerLite to measure the Cl—Cl bond distance in Figure 7.5c (also called the **intramolecular distance**). Then measure the *closest* distance* between two adjacent molecules of Cl_2 (the **intermolecular distance**). Repeat the activity for the orthorhombic iodine crystal in Figure 7.5b.

The Activity (in *Exploring the third dimension*, on one of the CD-ROMs) should take you approximately 15 minutes to complete.

Ice

Another example of a structure consisting of individual molecules held together in a three-dimensional arrangement is that of crystalline water or ice (Figure 7.6).

---

* You will find more than one intermolecular distance, and may have to hunt about for the shortest!

Figure 7.6
Crystals of ice.

● What is the shape of a water molecule?

○ Figure 7.7a gives the Lewis structure of water. Using VSEPR theory, two of the six electrons surrounding the central oxygen atom form two bonding pairs with the two electrons on the two hydrogen atoms, leaving two non-bonded electron pairs; hence there are four repulsion axes. The water molecule is thus predicted to be bent, with the bonding electron pairs at slightly less than the tetrahedral angle to each other (Figure 7.7b). This indeed is what is found experimentally for free gaseous molecules of water.

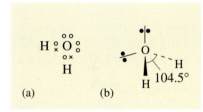

(a)   (b)

Figure 7.7
(a) Lewis structure of H_2O;
(b) repulsion axes of the H_2O molecule.

The structure of one of the polymorphs of ice is shown in Figure 7.8a; this is the hexagonal form (known as I_h), which forms at atmospheric pressure. Each water molecule in the crystal is coordinated tetrahedrally by four others.

● What kind of bonding do we find in crystals of ice?

○ The H_2O molecules have strong covalent bonds. The crystal is held together by hydrogen-bonding, where a hydrogen, $H^{\delta+}$, from one H_2O molecule is attracted to the electronegative, $O^{\delta-}$, atom of a neighbouring molecule; this is a special case of dipole–dipole interaction.

The hydrogen bonds create a continuous three-dimensional network. In general, the solid phase of any compound is denser than the liquid, but this open hydrogen-bonded network of water molecules makes ice less dense than water, so that it floats on the surface of water (Figure 7.8b). The extent of the hydrogen-bonding diminishes with increasing temperature, resulting in a maximum density of water at 4 °C, which has a significant impact on the circulation of water in the oceans, and hence our weather.

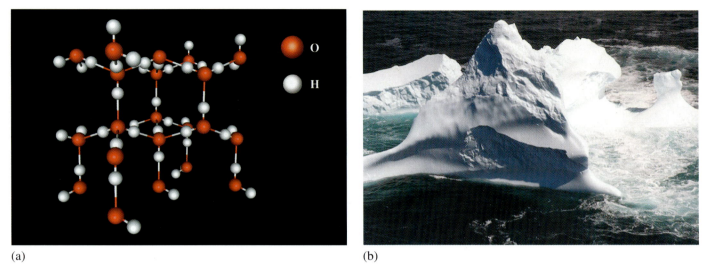

(a)   (b)

Figure 7.8 (a) The crystal structure of hexagonal ice (I_h); (b) a floating iceberg. 💻

In the Group VI series of compounds H_2Te, H_2Se and H_2S, which are all gases at room temperature, the boiling temperature falls progressively with decreasing molecular mass: 271 K, 232 K, and 212 K, respectively. (You may have come across the very poisonous and smelly gas, H_2S — hydrogen sulfide — which used to be generated in stink bombs before they became illegal, and can be smelt near volcanoes and hot springs.) This trend is sharply reversed in H_2O, which is a liquid at room temperature, boiling at 373 K. The anomalous boiling temperature of H_2O is another property that is explained by the phenomenon of hydrogen-bonding.

Carbon dioxide

Carbon dioxide provides us with an example of another triatomic covalent molecule, which, when cooled sufficiently, forms a molecular crystalline solid. As 'dry-ice', it is commonly used in laboratories for cooling purposes, and sometimes in the theatre to make smoke or misty effects.

🔵 What is the shape of a CO_2 molecule?

⚪ The four valence electrons of carbon are all used in bonding pairs to give two double bonds to the oxygens (Figure 7.9a). There are two repulsion axes, giving a linear molecule (Figure 7.9b).

Figure 7.10 illustrates the unit cell (and the corresponding packing diagram) of the carbon dioxide crystal structure. It is very easy to see that it consists of discrete linear CO_2 molecules, held together in the crystal by van der Waals forces.

Finally, we move to the structures of some organic molecules. Benzene is a molecule that you will have met before. For many years, its structure was the subject of debate. Luckily it crystallizes just below room temperature; indeed on cold days in winter, it was not uncommon in poorly heated laboratories to find that it had solidified. Hence it was possible for Kathleen Lonsdale (see Box 7.1) to apply X-ray crystallography to benzene crystals to confirm that its molecule was indeed a flat six-membered ring (Figure 7.11a and b). The crystal structure of a related molecule, naphthalene, is shown in Figure 7.11c.

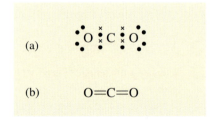

Figure 7.9
(a) Lewis structure of CO_2;
(b) shape of the CO_2 molecule.

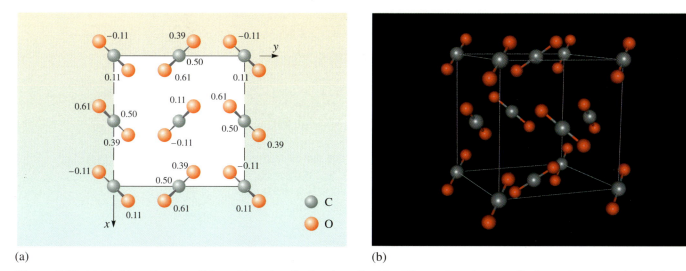

(a)

(b)

Figure 7.10 (a) Packing diagram of the cubic unit cell of carbon dioxide, CO_2, projected perpendicular to the z axis; the heights of the atoms are expressed in terms of the fractional coordinate c; (b) the unit cell of the CO_2 crystal structure.

The characteristics of the molecular crystals that we have discussed are that discrete molecules are easily discernible in the structure, and that they are held together by weak forces. As the intermolecular binding forces are weak, these crystals tend to have low melting and boiling temperatures.

Figure 7.11

The structure of benzene: (a) face-on and (b) side-on; (c) the crystal structure of naphthalene.

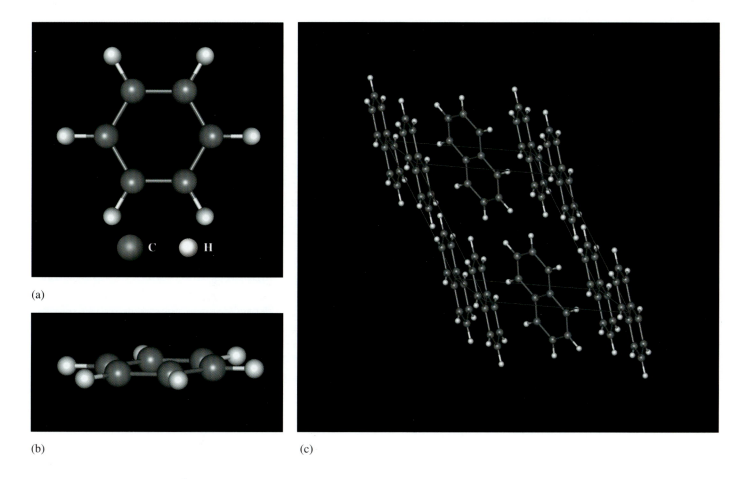

(a)

(b)

(c)

BOX 7.1 Dame Kathleen Lonsdale 1903–1971

Kathleen Lonsdale (Figure 7.12) was born in Newbridge, Ireland. After a brilliant undergraduate career, she started her research work in crystallography in 1922 with William Bragg at University College London. In 1945, she became one of the first women to be admitted as a Fellow of the Royal Society, and she became a professor in chemistry at UCL in 1946. A physicist and mathematician by training, much of her work lay with the theory of space groups and symmetry, but in her experimental work she was the first person to demonstrate that the benzene ring is hexagonal and planar. She also had a lifelong interest in the structure of diamond, and measured its carbon–carbon distance very accurately. She was a member of the Society of Friends (Quakers), and also worked for prison reform.

Figure 7.12
Dame Kathleen Lonsdale (1903–1971).

A summary of the types of crystalline solid that we have discussed is given in Table 7.3, which relates the type of structure to the physical properties of the solid. These descriptions are intended only as guidelines, as not all crystals will fit exactly into one category or another.

Table 7.3 Classification of crystal structures

Type	Structural unit	Bonding	Characteristics	Examples
ionic	cations and anions	electrostatic, non-directional	hard, brittle, crystals of high m.t.; moderate insulators; melts are electrically conducting	alkali metal halides
extended covalent array	atoms	mainly covalent	strong hard crystals of high m.t.; insulators	diamond, silica
molecular	molecules	mainly covalent between atoms in the molecule, van der Waals or hydrogen-bonding between molecules	soft crystals of low m.t. and large coefficient of thermal expansion; insulators	ice, organic compounds
metallic	metal atoms	cations and electrons	single crystals are soft; strength depends on structural defects and grain; good electrical conductors; their m.t.s tend to be high	iron, aluminium, sodium

7.4 Summary of Section 7

1 Molecular crystals are characterized by low melting temperatures and contain discrete, distinguishable molecules.

2 Weak van der Waals bonding or hydrogen-bonding holds molecular crystals together.

3 Cl_2, Br_2, and I_2 form orthorhombic crystals containing discrete diatomic molecules.

4 In crystalline water (ice) each water molecule is coordinated to four others by hydrogen-bonding.

5 CO_2 and benzene crystals contain discrete molecules held together by van der Waals forces.

QUESTION 7.1

What kind of van der Waals forces hold crystalline carbon dioxide together?

QUESTION 7.2

Figure 7.13 shows two schematic electron density maps for two different crystalline solids. What principal types of bonding do you think are present in each of these structures?

QUESTION 7.3

What are the principal interactions holding together each of the following: (i) ice; (ii) solid carbon monoxide, CO; (iii) potassium bromide, KBr; (iv) frozen nitrogen, N_2(s); (v) platinum.

QUESTION 7.4

By how much does the energy of interaction of the London dispersion forces decrease when the distance between two molecules doubles in size?

COMPUTER ACTIVITY 7.2

Use WebLab ViewerLite to determine the carbon–oxygen bond distance in crystalline CO_2, and the intermolecular oxygen–oxygen distance.

The Activity (in *Exploring the third dimension* on one of the CD-ROMs) should take you approximately 10 minutes to complete.

COMPUTER ACTIVITY 7.3

Use WebLab ViewerLite to determine the ∠H—O—H bond angle, the O—H covalent bond length, and the intermolecular H—O hydrogen-bonded distance in the I_h form of crystalline H_2O.

The Activity (in *Exploring the third dimension* on one of the CD-ROMs) should take you approximately 10 minutes to complete.

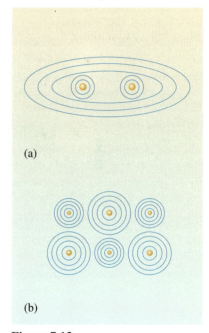

(a)

(b)

Figure 7.13
Schematic electron density maps for two different crystalline solids.

DEFECTS IN CRYSTALS

8

So far we have discussed crystals as though they were perfect structures. In practice, perfect crystals are never found.

Imperfections in crystals can arise during their growth; these are due to various types of dislocation, such as lines or layers of atoms being out of position in the lattice, and are known as *extended defects*. The strength of a material depends very much on the presence (or absence) of extended defects such as edge and screw dislocations and grain boundaries; these types of phenomena lie very much in the realm of materials science and will not be discussed here.

In a perfect crystal, all the atoms would be at their correct lattice positions in the structure. This situation can exist only at the absolute zero of temperature, 0 K. Above 0 K, defects occur in the structure at particular points within the lattice; these are known as **point defects**, and can be due to the presence of a foreign atom at a particular site or to the presence of a vacancy where normally one would expect an atom. Point defects can have significant effects on the chemical and physical properties of the solid. Examples are: (i) the beautiful colours of many gemstones such as ruby and sapphire are due to impurity atoms in the crystal structure; (ii) some ionic solids are able to conduct electricity by a mechanism that depends on the presence of vacant ion sites within the lattice.

Molecules are usually formed with exact integer ratios of atoms, expressed by formulae such as CO_2 and CH_4. This is a consequence of the valency possessed by atoms, and such compounds are said to be *stoichiometric*. However, in solid compounds this stoichiometry can be changed by the presence of defects.

Defects fall into two main categories: **intrinsic defects**, which are integral to the crystal in question (they do not change the overall composition, and because of this they are also known as **stoichiometric defects**); and **extrinsic defects**, which are created when an atom of another element is inserted into the lattice.

8.1 Stoichiometric defects

Stoichiometric defects in ionic structures occur where ions are either completely missing or are out of position, but where the overall stoichiometry is preserved.

Defects arising from **vacant sites** within the lattice are known as **Schottky defects**. This phenomenon is illustrated in Figure 8.1a for a layer through an NaCl-type crystal structure; if you compare this with the ideal situation (shown in Figure 8.1b), you can see that for every vacant cation position there is also a vacant anion site. A Schottky defect consists of this *pair of vacancies*, which thus preserves electrical neutrality and the overall stoichiometry.

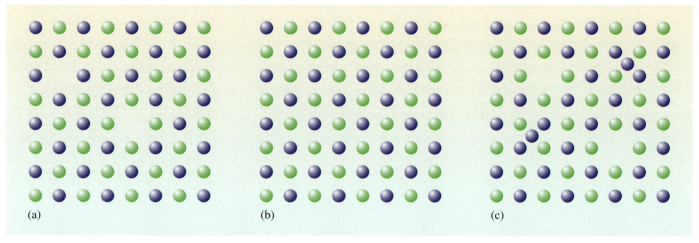

Figure 8.1 Point defects in an MX-type crystal structure: (a) Schottky defects; (b) perfect crystal; (c) Frenkel defects. The purple spheres represent cations and the green spheres represent anions.

In a $CaCl_2$ structure with n vacant Ca^{2+} sites, how many vacant Cl^- sites would you expect?

$2n$. A Schottky defect in $CaCl_2$ consists of three vacancies: for every vacant Ca^{2+} site there will be two vacant Cl^- sites.

A second type of defect occurs when an ion is displaced to occupy an **interstitial site** (that is, *not* at a lattice site), as in Figure 8.1c. These **Frenkel defects** occur more often in structures where the cation is much smaller than the anion.

Why are point defects important? Well, they provide a mechanism whereby atoms and ions can move or diffuse within a crystal; in particular, they have an application in photography (Box 8.1).

How could vacant sites be deliberately introduced into an ionic crystal?

If you introduce a small amount of, say, $MgCl_2$ into a NaCl lattice, each Mg^{2+} ion will occupy only one site for every two that would have been occupied by Na^+ ions, in order to preserve electrical neutrality. This can be important in producing changes in diffusion rates and in ionic conductivities in these crystals.

BOX 8.1 Crystals in action 9: the black-and-white photographic process

A photographic emulsion consists of very small crystals of AgBr (or AgBr–AgI) dispersed in gelatin: this is usually supported on paper or thin plastic to form film (Figure 8.2). The crystals are usually small triangular or hexagonal platelets, known as grains. They are grown very carefully *in situ* with as few structural defects as possible, their size in the range $0.05–2 \times 10^{-6}$ m. During the photographic process, light falling on the AgBr produces Ag atoms in some of the grains; these eventually form the dark parts of the negative. The grains that are affected by the light contain the so-called latent image. It is important for the grains to be free from structural defects, such as dislocations and grain boundaries, because these interfere with the deposition of the Ag atoms on the surface of the grains. However, the formation of the 'latent' image is dependent on the *presence* of point defects.

Figure 8.2 A black and white photograph and its negative.

AgBr has the NaCl crystal structure. However, unlike the alkali halides, which contain mainly Schottky defects, AgBr has been shown to contain mostly Frenkel defects, in the form of interstitial Ag^+ cations. For a grain to possess a latent image, it need have as few as *four* Ag atoms in a cluster on the surface. The formation of the clusters of Ag atoms is a complex process that is still not fully understood. However, it is thought to take place in several stages. The first stage is when light strikes one of the AgBr crystals and excites an electron (not a valence electron) from the core of an atom. The light absorbed is from the extreme blue end of the visible spectrum. This electron eventually neutralizes one of the interstitial silver cations, Ag_i^+:

$$Ag_i^+ + e^- = Ag$$

In the next stages, this Ag atom speck has to grow into a cluster of atoms on the surface of the crystal. A possible sequence of events for this is that the Ag atom interacts with another excited electron:

$$Ag + e^- = Ag^-$$

This silver anion could then neutralize an interstitial silver cation:

$$Ag^- + Ag_i^+ = Ag_2$$

and gradually a small cluster builds up:

$$Ag_2 + Ag_i^+ = Ag_3^+$$
$$Ag_3^+ + e^- = Ag_3$$
$$Ag_3 + e^- = Ag_3^-$$
$$Ag_3^- + Ag_i^+ = Ag_4, \text{ and so on}$$

It appears that only the odd-numbered clusters seem to interact with the electrons.

In reality, the process is even more complex than this because photographic emulsions made from pure AgBr are not sensitive enough. Sensitizers, such as sulfur or organic dyes, are added, which absorb light of longer wavelength than AgBr and so extend the spectral range. The sensitizers form traps for the electrons on the surfaces of the grains; these electrons then transfer from an excited energy level of the sensitizer to the AgBr.

The film containing the latent image is treated with various chemicals to produce a lasting negative. First of all it is developed: a reducing agent such as an alkaline solution of hydroquinone is used to reduce the AgBr crystals to Ag. The clusters of Ag atoms act as a catalyst to this reduction process, and all the grains with a latent image are reduced to Ag. The process is rate controlled, so the grains that have not reacted with the light are unaffected unless the film is developed for a very long time, when eventually they will be reduced and a fogged picture results. The final stage in producing a negative is to dissolve out the remaining light-sensitive AgBr: this is done using sodium thiosulfate, $Na_2S_2O_3$ (commonly known as hypo), which forms a water-soluble complex with Ag^+ ions. The dark areas on the final negative thus consist of the Ag formed in the grains with the latent images.

8.1.1 Colour centres

During the 1930s, in Germany, it was noticed that crystals of the alkali metal halides that had been exposed to X-rays became brightly coloured. The colour was thought to be associated with a defect known then as a Farbenzentrum (German *Farbe,* colour), now abbreviated to **F-centre**. Since then, it has been found that many forms of high-energy radiation (ultraviolet, X-rays, neutrons) will cause F-centres to form. The colour produced by the F-centres is always characteristic of the host crystal, so, NaCl for instance, becomes deep yellow–orange, KCl violet, and KBr blue–green.

Subsequently, it was found that F-centres can also be produced by heating a crystal of an alkali halide in the vapour of an alkali metal: this gives a clue to the nature of these defects. The excess alkali metal atoms diffuse into the crystal and settle at cation sites; at the same time, an equivalent number of anion site vacancies are created, and ionization gives alkali metal cations with electrons trapped at anion vacancies (Figure 8.3). In fact, it doesn't even matter which alkali metal is used: if NaCl is heated with potassium, the colour of the F-centre does not change because the colour is characteristic of the energy levels of the electrons trapped at the anion vacancies in the host halide.

Spectroscopic work has confirmed that F-centres are indeed unpaired electrons trapped at vacant anion sites. A series of energy levels are available for the electron, and the energy required to transfer from one level to another falls into the visible part of the electromagnetic spectrum — hence the colour of F-centres.

There is an interesting natural example of this phenomenon: the mineral Blue John (Figures 4.12 and 4.13) owes its beautiful blue–purple coloration to the presence of F-centres.

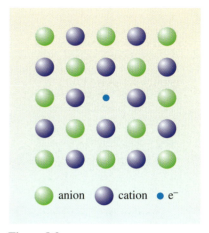

anion   cation e^-

Figure 8.3
The F-centre, an electron trapped at a vacant anion site in an alkali halide crystal.

8.2 Non-stoichiometric compounds

Many well-known oxides and sulfides do not have exact empirical formulae, and are known as **non-stoichiometric compounds**. This may be due to either a deficiency or an excess of metal atoms or ions. We shall just give one example of how this can work.

Metal deficiency

Consider the formula for iron(II) oxide. It is not FeO, as you might expect, but approximately $Fe_{0.95}O$. The compound has the NaCl structure, based on a *ccp* array of oxide ions, with the octahedral holes occupied by Fe^{2+} ions. However, some of the octahedral holes are vacant, whereas others, to preserve electrical neutrality, are occupied by Fe^{3+} ions, thus giving a compound with fewer iron atoms than oxygen atoms. Fe^{3+} ions are also found in interstitial tetrahedral sites.

In this Section, we have shown you, very briefly, that the solid state is not quite as perfect as we led you to believe initially. In fact, some of the most important properties and uses of solids in technology arise because of their imperfect nature.

8.3 Summary of Section 8

1   Ionic crystals contain two types of stoichiometric defects: Schottky (vacancy) and Frenkel (interstitial).

2   Colour centres can be created in alkali halide crystals; they are due to electrons trapped at vacant anion sites.

3   In non-stoichiometric compounds the charge balance can be maintained by variable valency metal ions.

QUESTION 8.1

Is a Schottky defect intrinsic or extrinsic?

QUESTION 8.2

Zirconia, ZrO_2 can be doped with small amounts of calcium. How will this affect the charge balance of the compound? Would you classify the defects in this structure as extrinsic or intrinsic?

SUMMARY OF *THE THIRD DIMENSION: CRYSTALS*

9

The emphasis in this part of the Book has been on the structure of crystalline solids. In Section 2, we concentrated on the ways in which spheres can pack together. The exercises involving model building provided a good analogy for the structures adopted by the common metals.

With the notion of arrays of identical atoms in mind, we moved on to examine extended structures in two- and three-dimensional space. We defined what is meant by a lattice, and introduced the basic building block of crystals, the unit cell. After this preliminary overview, we were able to move to real crystalline solids in Sections 4, 6 and 7. It is obviously impossible to mention the vast range of crystalline structures in a book of this length. However, the major classifications of metallic, ionic, extended covalent and molecular structures are useful in categorizing most solids.

All the discussion in Sections 1–4 implicitly assumed that ions and atoms could be regarded as incompressible spheres. Section 5 examined the validity of this concept, and explored ways of determining a set of self-consistent data for ionic radii.

In Part 2 of this Book we shall concentrate on the structure of individual organic molecules.

LEARNING OUTCOMES

Now that you have completed *The Third Dimension: Crystals*, you should be able to do the following things:

1 Recognize valid definitions of, and use in a correct context, the terms, concepts and principles in the following Table. (All Questions)

List of scientific terms, concepts and principles introduced in *Crystals*

Term	Page number	Term	Page number
anion	46	dipole–dipole interaction	90
antifluorite structure	58	effective ionic radius	74
ball-and-stick model	24	electric dipole	90
body-centred cubic (*bcc*)	24	extrinsic defect	100
body-centred lattice, I	40	F-centre	103
cadmium chloride structure	59	face-centred lattice, F	40
cadmium iodide structure	59	fluorite structure	56
caesium chloride structure	47	fractional coordinates	48
carbon dioxide structure	96	Frenkel defect	101
cation	46	graphite structure	87
centred unit cell	37	hexagonal close-packing (*hcp*)	21
charge–dipole interaction	90	high-temperature superconductor	64
close-packed structure	16	hydrogen-bonding	91
coordination number	24	intermolecular distance	94
corundum structure	61	interstitial site	101
Coulomb's law	46	intramolecular distance	94
counter ion	74	intrinsic defect	100
covalency	46	inverse square law	46
covalent bond	90	ionic bond	46
covalent radius	92	ionic radius	69
crystal lattice	32	isoelectronic species	71
crystal radius	74	lattice point	32
cubic close-packing (*ccp*)	22	layer structure	80
cubic unit cell	39	London dispersion force	91
diamond	82	metallic bonding	89

2   Use simple geometry to determine distances between objects in two and three dimensions. (Questions 2.3 and 2.4)

3   Given a diagram of two close-packed layers of spheres, describe the different possibilities for the third layer in terms of cubic and hexagonal close-packing. (Question 4.16)

4   Understand the number, relative sizes and positions of octahedral and tetrahedral holes in close-packed structures. (Questions 2.1, 2.6, 4.9, 4.13, 4.14, 4.17, 4.20 and 4.21)

5   Describe the coordination of the atoms in some simple crystal structures and recognize their unit cells. (Questions 2.2, 2.5, 2.6, 3.3, 4.1, 4.3, 4.9, 4.10, 4.13, 4.14, 4.17, 4.20 and 4.21)

6   Given the unit cell of a crystalline compound, calculate the number of molecules in it. (Questions 3.1, 3.2, 3.4, 4.2, 4.4 and 4.12)

7   From a knowledge of the size and contents of a unit cell, calculate the density of a crystal. (Questions 3.2, 4.2, 4.4, 4.6, 4.7, 4.8 and 4.12)

8   Describe the structures of some simple MX_n ionic solids in terms of the occupation of octahedral and tetrahedral holes in close-packed structures. (Questions 4.3, 4.9, 4.10, 4.13, 4.14, 4.16, 4.17, 4.20, 4.21 and 4.22)

9   Draw projections (packing diagrams) of the unit cells of simple crystal structures. (Questions 4.5, 4.11, 4.15, 4.16, 4.18 and 4.19)

10 Calculate ionic radii from a table of internuclear distances of ionic compounds. (Questions 5.1 and 5.2)

11 Understand the different criteria that determine the stability of a crystal structure. (Questions 5.3 and 5.4)

12 Given a list of crystalline compounds, identify and describe the different types of bonding present. (Questions 6.1, 7.1, 7.2, 7.3 and 7.4)

13 Be able to use a crystallographic plotting package such as WebLab ViewerLite to measure inter- and intramolecular distances and bond angles. (Computer Activities 6.1, 6.2, 6.3, 7.1, 7.2 and 7.3)

14 Understand the difference between intrinsic and extrinsic defects in crystal structures. (Questions 8.1 and 8.2)

QUESTIONS: ANSWERS AND COMMENTS

QUESTION 2.1 (*Learning Outcome 4*)

There are $2n$ tetrahedral holes and n octahedral holes, the same as for an *hcp* array.

QUESTION 2.2 (*Learning Outcome 5*)

Twelve. Check this from the image in WebLab ViewerLite. Again, this is the same as for the *hcp* array.

QUESTION 2.3 (*Learning Outcome 2*)

(a) The triangles necessary to work out these distances are shown in Figure Q.1.

AB, BC and DC are all cube sides, with length a. Using the Pythagoras theorem:

$$AC^2 = AB^2 + BC^2 = a^2 + a^2 = 2a^2$$

$$AC = \sqrt{2}\,a$$

Similarly, it follows that

$$AD^2 = DC^2 + AC^2 = a^2 + 2a^2 = 3a^2$$

$$AD = \sqrt{3}\,a$$

Thus, x, the distance from the atom at the centre of the cell to an atom at a corner is half of AD:

$$x = \frac{\sqrt{3}}{2}\,a = 0.866a$$

(b) The distance from an atom at the centre of the cell to the atom at the equivalent position in the next cell is a (see Figure Q.1).

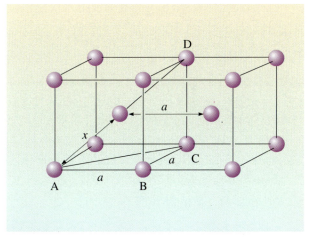

Figure Q.1
Right-angled triangles in the body-centred cubic structure.

QUESTION 2.4 (*Learning Outcome 2*)

The Pythagoras theorem tells us that in Figure 2.16 $a^2 + b^2 = c^2$. In this case, $a = 3$ and $c = 5$, and we want to know what b is, so

$$3^2 + b^2 = 5^2$$

Rearranging,

$$b^2 = 5^2 - 3^2 = 25 - 9 = 16$$

Thus, $b = \sqrt{16} = 4$.

QUESTION 2.5 (*Learning Outcome 5*)

From an inspection of the model in Figure 2.18, you should be able to see that each sphere would be in direct contact with six others at the corners of an octahedron: the coordination number is six, and the geometrical arrangement is octahedral (Figure Q.2).

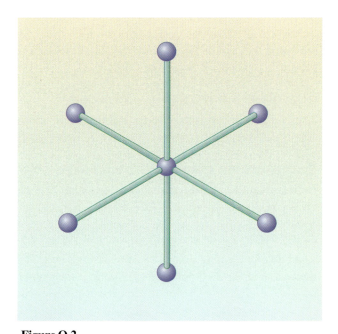

Figure Q.2
Model of the octahedral disposition of atoms around a central atom in an infinite primitive cubic array.

QUESTION 2.6 *(Learning Outcomes 4 and 5)*

An atom occupying a tetrahedral hole will be four coordinate, and an atom occupying an octahedral hole will be six coordinate.

QUESTION 3.1 *(Learning Outcome 6)*

P   The corner of each unit cell is shared with eight other unit cells, so only one-eighth of an atom is in each cell. The unit cell has eight corners and, therefore, contains one atom (Figure Q.3a).

I   The body-centred cell has ($8 \times 1/8$) atoms at the corners, and one atom in the centre (which is not shared), making two atoms in all (Figure Q.3b).

F   Each face is shared by two unit cells. Therefore the face-centred cell has ($6 \times \frac{1}{2}$) atoms on the faces and ($8 \times 1/8$) at the corners, making a total of four atoms (Figure Q.3c).

QUESTION 3.2 *(Learning Outcomes 6 and 7)*

A *bcc* unit cell contains two atoms of element X. The volume of a unit cell of element X is

$$(286.6 \times 10^{-12})^3 \, \text{m}^3 = 2.354 \times 10^{-29} \, \text{m}^3$$

The mass of a unit cell, or two atoms of X, will be

$$\text{density} \times \text{volume} = (7.874 \times 10^3 \times 2.354 \times 10^{-29}) \, \text{kg}$$

So, the mass of one mole of X will be

$$\tfrac{1}{2}(7.874 \times 10^3 \times 2.354 \times 10^{-29}) \times 6.022\,0 \times 10^{23} \times 10^3 \, \text{g} = 55.81 \, \text{g}$$

Checking the relative atomic mass table in the *Data Book* (from the CD-ROM), we see that element X must be iron.

QUESTION 3.3 *(Learning Outcome 5)*

There are only three unit cells marked on Figure 3.18. A and D are not unit cells, as repeating these units side by side would not give the original pattern. Unit cell C is primitive, whereas B and E are centred.

QUESTION 3.4 *(Learning Outcome 6)*

The unit cell of caesium is body-centred. There are therefore two atoms in the unit cell, one at the centre of the cell and ($8 \times 1/8$) = 1 at the corners.

QUESTION 4.1 *(Learning Outcome 5)*

Both Cs^+ and Cl^- have a coordination number of eight, as each ion is surrounded by eight equidistant oppositely charged ions at the corners of a cube.

QUESTION 4.2 *(Learning Outcomes 6 and 7)*

There is one formula unit of CsCl in the unit cell. This is made up from one Cs^+ in the centre of the cell, plus eight Cl^- ions at the corners. Each of the Cl^- ions contributes one-eighth of a Cl^- ion to the cell, as it is shared by eight adjacent unit cells.

QUESTION 4.3 *(Learning Outcomes 5 and 8)*

The coordination number is six for both Na^+ and Cl^- ions; each ion is octahedrally coordinated by the other.

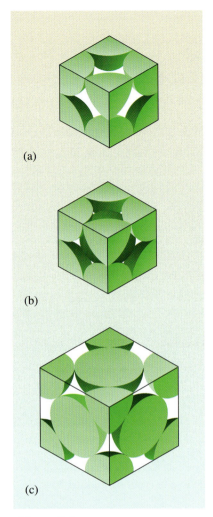

(a)

(b)

(c)

Figure Q.3
Unit cells showing occupation:
(a) primitive, (b) body-centred,
(c) face-centred.

QUESTION 4.4 (Learning Outcomes 6 and 7)

There are four Cl^- ions: eight at cube corners shared with eight unit cells, and six at cube faces shared between two unit cells, making a total of $(8 \times \frac{1}{8}) + (6 \times \frac{1}{2}) = 4$. There are four Na^+ ions: twelve at midpoints of cube edges shared with four unit cells, and one unshared at the body-centre position, making a total of $(12 \times \frac{1}{4}) + 1 = 4$. Thus, there are four formula units of NaCl in the unit cell. Figure Q.4 may make this easier to see.

QUESTION 4.5 (Learning Outcome 9)

The unit cell projection for NaCl is shown in Figure Q.5.

QUESTION 4.6 (Learning Outcome 7)

The total relative mass of the NaCl unit cell is

$(4Na + 4Cl) = (4 \times 22.989\,8 + 4 \times 35.453) = 233.77$

Converting this to grams, we get:

$$\frac{233.77\,g}{6.022\,0 \times 10^{23}} = 3.881\,9 \times 10^{-22}\,g$$

The volume of the unit cell is

$(564 \times 10^{-10}\,cm)^3 = 1.79 \times 10^{-22}\,cm^3$

The density is therefore

$$\frac{3.8819 \times 10^{-22}\,g}{1.79 \times 10^{-22}\,cm^3} = 2.17\,g\,cm^{-3}$$

(In SI units this becomes $2.17 \times 10^3\,kg\,m^{-3}$.)

QUESTION 4.7 (Learning Outcome 7)

The total relative mass of the AgCl unit cell is

$4Ag + 4Cl = (4 \times 107.868 + 4 \times 35.453) = 573.284$

Thus, the total mass of one unit cell will be

$$\frac{573.284\,g}{6.022\,0 \times 10^{23}} = 9.5198 \times 10^{-22}\,g$$

The volume of the unit cell will be given by

$$volume = \frac{mass}{density} = \frac{9.5198 \times 10^{-22}\,g}{5.571\,g\,cm^{-3}} = 1.709 \times 10^{-22}\,cm^3$$

The unit cell dimension is thus

$(1.709 \times 10^{-22}\,cm^3)^{1/3} = 5.549 \times 10^{-8}$ cm, or 554.9 pm.

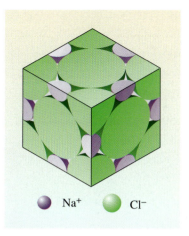

Na^+ Cl^-

Figure Q.4
NaCl unit cell, showing occupation by ions.

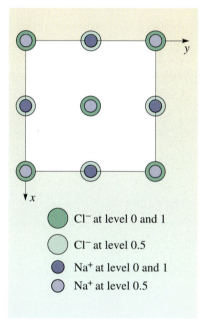

Cl^- at level 0 and 1

Cl^- at level 0.5

Na^+ at level 0 and 1

Na^+ at level 0.5

Figure Q.5
Packing diagram for the NaCl structure.

QUESTION 4.8 (Learning Outcome 7)

If we write the relative atomic mass of X as x, the total relative mass of the AgX unit cell is

$$4Ag + 4x = 4 \times 107.868 + 4x = 431.472 + 4x$$

Thus, the total mass of one unit cell will be

$$\frac{431.472 + 4x}{6.022\,0 \times 10^{23}}\ g$$

The volume of the unit cell will be

$$(577.5 \times 10^{-12})^3\ m^3 = 1.926 \times 10^{-28}\ m^3$$

The mass of the unit cell will be

$$\frac{431.472 + 4x}{6.022\,0 \times 10^{26}}\ kg = 6.477 \times 10^3\ kg\ m^{-3} \times 1.926 \times 10^{-28}\ m^3$$
$$= 1.247 \times 10^{-24}\ kg$$

Multiplying through by $6.022\,0 \times 10^{26}$, we find that

$$431.5 + 4x = 751.2$$
$$4x = 319.7$$

so, the relative atomic mass of X = 79.93. Element X is therefore bromine.

QUESTION 4.9 (Learning Outcomes 4, 5 and 8)

This is indeed possible, and it is the structure adopted by nickel arsenide, NiAs.

QUESTION 4.10 (Learning Outcomes 5 and 8)

The coordination number is four in each case. Each Zn^{2+} ion is surrounded tetrahedrally by four S^{2-} ions, and vice versa.

QUESTION 4.11 (Learning Outcome 9)

The packing diagram for ZnS (zinc blende) is shown in Figure Q.6, with sulfur at the corners of the unit cell.

QUESTION 4.12 (Learning Outcomes 6 and 7)

Four. For the zincs in Figure 4.10, there are $(6 \times \frac{1}{2}) = 3$ at the centres of the faces and $(8 \times \frac{1}{8}) = 1$ at the corners. These four are matched by the four sulfurs entirely enclosed in the cell.

QUESTION 4.13 (Learning Outcomes 4, 5 and 8)

The *wurtzite structure* is based on a hexagonal close-packed array of sulfur, with zinc ions occupying alternate tetrahedral holes (Figure Q.7).

Structures of the same element or compound that differ only in their atomic arrangements, such as the two forms of ZnS, are termed *polymorphs*.

Compounds that adopt the wurtzite structure include BeO, ZnO and NH_4F.

STUDY NOTE

How to construct the wurtzite structure is illustrated in 'The wurtzite structure' in *Model Building* on one of the CD-ROMs associated with this Book.

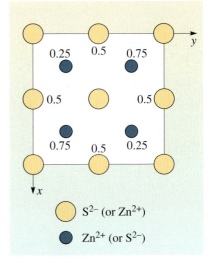

Figure Q.6
Packing diagram for the zinc blende, ZnS, structure.

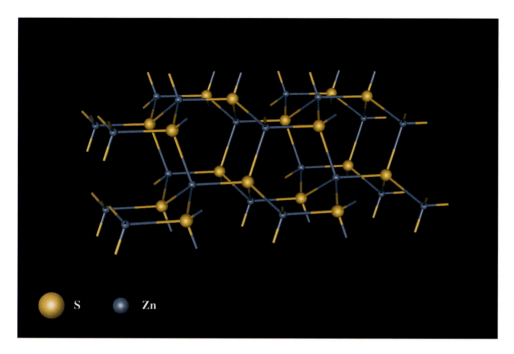

Figure Q.7
The crystal structure of wurtzite, ZnS. 💻

QUESTION 4.14 (Learning Outcomes 4, 5 and 8)

Each zinc atom is surrounded by four sulfur atoms arranged tetrahedrally, and vice versa.

QUESTION 4.15 (Learning Outcome 9)

The packing diagram for CaF_2 is shown in Figure Q.8.

QUESTION 4.16 (Learning Outcomes 3, 8 and 9)

(a) Figure Q.9a shows the packing diagram for the *hcp* unit cell.

(b) Figure Q.9b shows the packing diagram for a possible unit cell for *ccp*. A more usual unit cell for this structure is the cubic one, which is less easy to visualize.

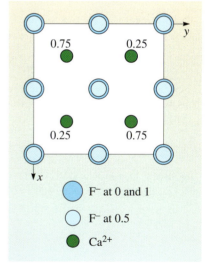

Figure Q.8
Packing diagram for the CaF_2 structure.

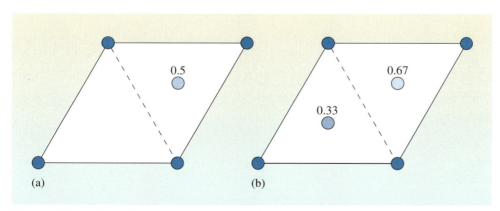

Figure Q.9 Packing diagrams for (a) hexagonal close-packing, and (b) cubic close-packing.

QUESTION 4.17 (Learning Outcomes 4, 5 and 8)

The coordination of the bismuth in this structure is octahedral (the bismuths are occupying octahedral holes). The formula is BiI_3, so if six iodines coordinate one bismuth atom, *two* bismuths must coordinate one iodine atom in order to maintain the correct ratio. In the structure, two bismuths are close to an iodine on one side, and on the other side there are iodines from the next layer.

QUESTION 4.18 (Learning Outcome 9)

Figure Q.10 shows the packing diagram for ReO_3.

QUESTION 4.19 (Learning Outcome 9)

Figure Q.11 shows the packing diagram for perovskite, ABX_3.

QUESTION 4.20 (Learning Outcomes 4, 5 and 8)

If there are n Cl^- ions, there are $2n$ tetrahedral holes, and so $2n/3$ are occupied by ions of M. The empirical formula of the compound is thus $M_{2n/3}Cl_n$. Cancelling n, and clearing fractions allows us to write this more conventionally as M_2Cl_3.

QUESTION 4.21 (Learning Outcomes 4, 5 and 8)

If there are n O^{2-} ions, there are n octahedral holes and so $n/2$ are occupied by ions of M. The empirical formula of the compound is thus $M_{n/2}O_n$. Cancelling n, and clearing fractions allows us to write this more conventionally as MO_2. Metal M thus has a valency of 4 in this compound.

QUESTION 4.22 (Learning Outcome 8)

There are four oxygens, so there will be $4 \times 2n = 8n$ tetrahedral holes. The n Mg^{2+} ions will occupy one-eighth of these. There are $4n$ octahedral holes, and the $2n$ Al^{3+} ions occupy half of these.

QUESTION 5.1 (Learning Outcome 10)

Assuming that anion–anion contact occurs as in Figure 5.2b, the radius of the iodide ion is OC, the value of which is obtained as follows:

$$\cos 45° = \frac{OC}{OA}$$
$$OC = OA \cos 45°$$
$$= \frac{302}{\sqrt 2} = 214 \, pm$$

QUESTION 5.2 (Learning Outcome 10)

From the internuclear distance in NaI,

$$r(Na^+) = (323 - 214) \, pm = 109 \, pm$$

Then, from the internuclear distance in NaF,

$$r(F^-) = (231 - 109) \, pm = 122 \, pm$$

The same procedure for RbI and RbF gives

$$r(Rb^+) = 152 \, pm \quad and \quad r(F^-) = 130 \, pm$$

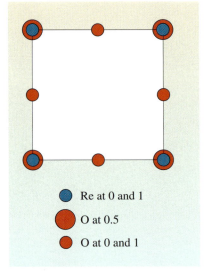

Figure Q.10
Packing diagram for ReO_3.

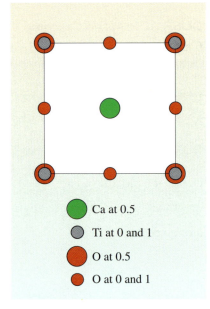

Figure Q.11
Packing diagram for perovskite, $CaTiO_3$ (ABX_3).

QUESTION 5.3 (Learning Outcome 11)

The most important factors are:

    (i)   the numbers of oppositely charged ions surrounding an ion;

    (ii)   the relative sizes of the ions;

    (iii) the polarizability of the ions;

    (iv) the charges on the ions.

QUESTION 5.4 (Learning Outcome 11)

Layer structures tend to be found when:

    (i)   there is only a small difference in electronegativity between the atoms;

    (ii)   when a large anion is polarized by a small cation.

QUESTION 6.1 (Learning Outcome 12)

The carbon–carbon distance in graphite is 142 pm, which lies between the values for single and double carbon–carbon bonds, and is very similar to the C–C bond length of 140 pm found in benzene. The network of linked hexagons in graphite has similar delocalized bonding to that found in benzene, where the bonding is halfway between single and double.

QUESTION 7.1 (Learning Outcome 12)

Carbon dioxide is a symmetrical linear molecule. The displacement of negative charge to the oxygen atoms exactly cancels each other out, so there is no resultant dipole. The forces between the molecules must therefore be due to induced dipole–induced dipole interactions or London dispersion forces.

QUESTION 7.2 (Learning Outcome 12)

Figure 7.13a shows contours typical of the electron density distribution between two atoms, with a build up of electron density between them; this is indicative of covalent bonding. van der Waals forces would thus be the only forces holding this crystal together. Figure 7.13b is indicative of an ionic lattice, as there is very low electron density between the atoms, and a spherical electron distribution around the atomic centres.

QUESTION 7.3 (Learning Outcome 12)

(i)   There are intramolecular covalent bonds in water molecules and hydrogen-bonding between the molecules.

(ii) CO has strong covalent bonds. The CO molecule is not symmetrical, so it possesses a small electric dipole. There will thus be van der Waals forces (dipole–dipole interactions) between the molecules.

(iii) KBr is an ionic solid with the same structure as NaCl.

(iv) N_2 molecules are symmetrical and possess no electric dipole, so crystalline N_2 will be held together by London dispersion forces (induced dipole interactions).

(v) Platinum is a metal whose atoms are held together by metallic bonding.

QUESTION 7.4 (Learning Outcome 12)

The energy of interaction decreases as $1/r^6$, so if r doubles to $2r$, the energy becomes $1/(2r)^6 = 1/64r^2$. Thus, the energy falls to 1/64th of its initial value.

QUESTION 8.1 (Learning Outcome 14)

Intrinsic. Schottky vacancies occur in pairs of cations and anions, maintaining charge neutrality and the stoichiometric ratio.

QUESTION 8.2 (Learning Outcome 14)

Calcium has a valency of +2, and in ZrO_2, zirconium has a valency of +4. If a zirconium is replaced by a calcium, then there will be a charge imbalance of −2 in the structure. The structure copes with this by eliminating an oxide ion, O^{2-}, leaving a vacancy. Such defects are termed *extrinsic*, as they have been deliberately created in the crystal.

ANSWERS TO COMPUTER ACTIVITIES

COMPUTER ACTIVITY 6.1 (*Learning Outcome 13*)

You should have found that all carbon–carbon distances are equal at 154 pm. The ∠C—C—C bond angle is 109.5°, the tetrahedral angle.

COMPUTER ACTIVITY 6.2 (*Learning Outcome 13*)

You should have found that in the β-quartz form shown here the Si–O distance is 162 pm. The linked SiO_4 tetrahedra are slightly distorted from regular symmetry with ∠O—Si—O— bond angles of 101.3°, 111.3° and 116.1°.

COMPUTER ACTIVITY 6.3 (*Learning Outcome 13*)

You should have found that the carbon–carbon distances within the layers are 142 pm, and the ∠C—C—C bond angles are 120°. The shortest carbon–carbon distance between the layers is 340 pm (Figure 6.9).

COMPUTER ACTIVITY 7.1 (*Learning Outcome 13*)

The van der Waals radius of chlorine is taken to be 175 pm. This means that if chlorine atoms in a crystal structure are closer than 350 pm (175 + 175), there is assumed to be some bonding between them. You should have found that for the chlorine crystal, the Cl—Cl bond distance is 197 pm, whereas the shortest non-bonded distance is 332 pm. The difference between bonded and non-bonded distances is therefore large, showing that there is very little interaction or bonding between chlorine molecules. The results for iodine are rather different. You should have found that the I—I bond distance is 271 pm, but that the shortest intermolecular I—I distance is only 350 pm. If there were no interaction between neighbouring I_2 molecules, we would have expected this value to be 396 pm, so clearly there is additional bonding between molecules. (This is discussed further in *Elements of the p Block*[2].)

COMPUTER ACTIVITY 7.2 (*Learning Outcome 13*)

The C=O bond distance is 116 pm, and the shortest intermolecular non-bonded O—O distance is 318 pm. If you measured the angle, you should have found the expected 180° for a linear molecule.

COMPUTER ACTIVITY 7.3 (*Learning Outcome 13*)

The O—H bond length is 101 pm, the intermolecular hydrogen-bonded H—O distance is 175 pm, and the ∠H—O—H bond angle is 109.5°. Notice that the symmetry of the ice crystal structure is now constraining the ∠H—O—H to be regular tetrahedral, whereas in a free molecule of water it is 104.5°.

FURTHER READING

1   A. Northedge, J. Thomas, A. Lane and A. Peasgood, *The Sciences Good Study Guide*, The Open University (1997).

2   C. J. Harding, R. Janes and D. A. Johnson (eds), *Elements of the p Block*, The Open University and the Royal Society of Chemistry (2002).

3   E. A. Moore (ed.), *Molecular Modelling and Bonding*, The Open University and the Royal Society of Chemistry (2002).

ACKNOWLEDGEMENTS

Grateful acknowledgement is made to the following sources for permission to reproduce material in Part 1 of this Book:

Text

p. 86: 'Misalliance' from *At the Drop of a Hat*, used by permission of The Flanders & Swann Estates © 1955. From *The Songs of Michael Flanders & Donald Swann.* IMP-Warner/Chappell (1996).

Figures

Figure 1.1c: courtesy of Diamond Trading Company; *Figure 1.2a*: Andrew Syred/Science Photo Library; *Figure 2.2b*: TM Microscopes/Vecco Metrology Group; *Figure 2.2c*: © National Power; *Figure 2.3a*: Dr Mitsuo Ohtsuki/ Science Photo Library; *Figure 2.3b*: courtesy of Topometrix Corporation; *Figures 2.4a and Figure 2.4b*: from *Chemistry and Chemical Reactivity,* 2nd edn, by John C. Kotz and Keity F. Purcell, copyright © by Saunders College Publishing, reproduced by permission of the publisher; *Figure 2.9:* Astrid and Hans-Frieder Michler/ Science Photo Library; *Figure 2.12a*: courtesy of The Royal Institution; *Figure 2.12b*: courtesy of Naomi L. Over and Antique Publications; *Figure 3.2b*: courtesy of St. Helen's, Bishopsgate, London; *Figure 3.4*: © Kongelige Bibliotek, Copenhagen; *Figure 3.8a and Figure 3.8b*: from *The Double Helix* by J. D. Watson, Weidenfeld & Nicolson (1968); *Figure 3.8c*: © Cold Spring Harbor Laboratory Archives; *Figure 3.10*: © The Nobel Foundation; *Figure 4.2:* Attard's Minerals, San Diego; *Figure 4.4c*: courtesy of the Wieliczka Salt Mines, Poland; *Figure 4.4d*: Simon Fraser/SPL; *Figure 4.5a*: Dr Jeremy Burgess/Science Photo Library; *Figure 4.9*: Ben Johnson/Science Photo Library; *Figure 4.12*: J. C. Revy/Science Photo Library; *Figure 4.13*: The Institute of Geological Sciences; *Figure 4.17a*: Roberto de Gugliemo/Science Photo Library; *Figure 4.17b*: Dr J. M. Whitehead; *Figure 4.17c*: courtesy of Dulux (Weathershield range, Brilliant White smooth masonry paint); *Figure 4.21a*: courtesy of P. P. Edwards; *Figure 4.21b*: Science Museum/Science and Society Picture Library; *Figure 4.22*: Texas Center for Superconductivity at the University of Houston; *Figure 4.25*: Scripps Research Institute; *Figure 5.1*: H. Schoknecht (1955) *Zeitschrift für Physikalische Chemie*, Vol. 3, R. Oldenbourgh Verlag GmbH; *Figure 5.5*: © The Nobel Foundation; *Figure 6.1*: E. R. Degginger/Science Photo Library; *Figure 6.3*: General Electric Company; *Figure 6.4*: © The Institute of Geological Sciences; *Figure 6.5:* Sinclair Stammers/Science Photo Library; *Figure 6.7c*: courtesy of Bacchus Antiques; *Figure 6.7d*: courtesy of Gene Buffalo; *Figure 7.6*: Dr Jeremy Burgess/Science Photo Library; *Figure 7.8b*: courtesy of International Ice Patrol; *Figure 7.12*: Godfrey Argent Studio, by courtesy of the National Portrait Gallery.

We would also like to thank Dr A. Tindle of the Open University Earth and Planetary Sciences Department for Figures 1.3b, 3.1, 3.4a and b, and 4.4a, and the cover image.

Part 2

Molecular shape

edited by Michael Gagan

*based on The Shapes of Molecules,
by Alan Bassindale (1991)*

THE TETRAHEDRAL CARBON ATOM

1

In molecular crystals, we have seen how the molecules of a solid compound are held together in a crystal array by weak intermolecular forces, such as hydrogen-bonding and London dispersion forces, rather than the strong electrostatic forces present in ionic crystals. We now need to examine what determines the shape of the individual molecules of the many carbon compounds that make up the realm of organic chemistry.

The isolated carbon atom has four electrons in its outer shell, and therefore, in combination with other elements, each neutral carbon nucleus will be surrounded by four pairs of electrons.

⚫ Using VSEPR theory, what will be the shape of a compound that has four electron pairs, or four single bonds, distributed around a central atom?

⚫ In order that the four electron pairs might be as far away from each other as possible, the structure will need to have its four bonds directed to the four corners of a tetrahedron. This is shown in Figure 1.1 for the methane, CH_4, molecule. It has also been shown experimentally that the angle between any two of the four bonds to the central carbon atom is exactly tetrahedral (109° 28′).

In this second Part of *The Third Dimension* we shall be concentrating on carbon compounds, but first a slight diversion on compounds containing nitrogen or oxygen is appropriate. The geometry around an oxygen atom in water, for example, involving two single O—H bonds is 'V-shaped', and around a nitrogen atom in ammonia involving three single —N—H bonds it is pyramidal.

The reason for this is that oxygen in water (and also in organic molecules containing single C—O bonds, like ethers and alcohols) is surrounded by four pairs of electrons — two pairs of bonding electrons, and two non-bonding electron pairs — with each pair taking up a different position in the tetrahedral array. Nitrogen in ammonia and amines also has four bonding pairs, but here there are three bonding pairs and one non-bonding pair. So, water and ammonia have the structures shown in Structures **1.1** and **1.2**[*], respectively.

⚫ What angle would you expect to find between the two oxygen–hydrogen bonds in water; and between two of the nitrogen–hydrogen bonds in ammonia?

⚫ Both angles are found to be less than the tetrahedral angle, because of the stronger repulsion from a non-bonded electron pair than from a bonding pair. In water the angle is 104.5°, and in ammonia it is 107.8°.

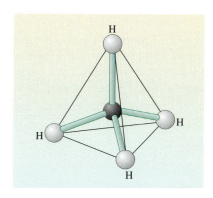

Figure 1.1
Model of a methane molecule. (Drawings of this and succeeding 'models' are based on the atom centres and straws of the Orbit model kit; see Box 1.2.)

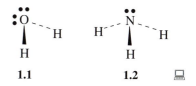

1.1   1.2

---

[*] Remember that the symbol, 🖥, indicates that this Figure is available in WebLab ViewerLite on one of the CD-ROMs associated with this Book.

We can look at this another way. Atoms in a molecule that are not directly bonded to each other are said to be **non-bonded**. Non-bonded atoms repel one another if they come too close together, as a result of the repulsion between the electron clouds surrounding each atom. Repulsion increases dramatically as the atoms get closer and closer together.

This approach also implies that the four atoms around a central carbon atom tend to be arranged so that they are as far apart as possible within the limits of fixed bond lengths. So in any molecule, CX_4, the four X groups will be located at the corners of a regular tetrahedron, with the carbon atom at the centre, since this is the arrangement that allows the four atoms bonded to the central atom to be as far apart as possible. If one of the attached atoms or groups is larger than the others, as in bromomethane, CH_3Br, the structure is still based on a tetrahedron. However, the large atom or group (here, bromine) has the effect of 'squeezing' the others closer together, thereby reducing the tetrahedral angle between the smaller atoms or groups. The size of an atom or group is determined by its surrounding electron cloud, and for atoms this is measured as its van der Waals radius (given in Table 1.1 in comparison with its covalent radius; this is part of the data in Table 7.2 (p. 93) of Part 1, where the covalent radii were given first). Table 1.1 includes values for those atoms commonly found in organic molecules. A fuller list is also given in the *Data Book* from one of the CD-ROMs associated with this Book.

Table 1.1 van der Waals (and single-bond covalent) radii for the common atoms in organic molecules

Atom	Radius/pm	Atom	Radius/pm
C	170 (77)	F	147 (71)
H	120 (37)	Cl	175 (99)
O	152 (73)	Br	185 (114)
N	155 (74)	I	198 (133)
S	180 (104)		

We have introduced this alternative description because non-bonded interactions between atoms in molecules have a number of important consequences in organic chemistry.

This concept of the **tetrahedral carbon atom** is so fundamental to our present understanding of structural organic chemistry, that it is difficult to believe that until 1874 (and in some cases beyond), chemists considered molecular shape to be of little importance, or else believed that it would never be possible to obtain information about the physical structure of molecules. Molecules, and the formulae that were used to represent them, were simply a useful abstract idea that enabled chemists to describe and systematize chemical reactions.

> The creed of the organic chemists of the first half of the 19th Century, that a chemical formula could not, and indeed should not represent more than an epitome of reactions, condemned as heretical any more imaginative speculations.
>
> W. G. Palmer, *A History of the Concept of Valency to 1930* (1965)

In 1874, J. H. van't Hoff (Figure 1.2) and J. A. Le Bel independently suggested that molecules had 'body'. The very simple, logical and elegant proposal, that any four groups around a carbon atom should be tetrahedrally disposed, is due to these two chemists. Organic chemistry was about to change.

Although van't Hoff and Le Bel's suggestion was received with scorn and abuse by the chemical establishment (Box 1.1), it soon became apparent that application of this simple idea could add a whole new dimension to organic chemistry. The premise of the tetrahedral carbon atom allows many observations of organic chemical behaviour to be explained. Their proposal was not based on theoretical considerations — they knew nothing about electrons and their involvement in bonding — but on practical observation. Using your model kit, you will be able to follow their line of argument (see Box 1.2, p. 125).

Figure 1.2
J. H. van't Hoff (1852–1911) was a Dutch chemist, who, with Le Bel, discovered the tetrahedral nature of the carbon atom. He was one of the founders of physical chemistry. After studying under Kekulé, he became a lecturer at the Veterinary College at Utrecht (1876), later succeeding to the Chairs of Chemistry in Amsterdam and Berlin. His research in physical chemistry included the theory of solutions, reaction kinetics, solid solutions, thermodynamics and the phase rule. In 1901, he was awarded the first Nobel Prize for Chemistry.

MODEL EXERCISE 1.1
The arrangement of bonds to a carbon atom

This exercise is designed to show you a little of the thinking that led to the idea of the tetrahedral carbon atom. Before 1874, methane and its derivatives might have been thought to be square planar (if they were thought to have a shape at all) — that is, all five atoms in a plane with 90° angles between adjacent bonds. It has been known for well over 100 years that no matter what experiments are tried, and however hard one searches, there is only one compound of formula CH_2Cl_2. Use your model kit to make as many different arrangements of CH_2Cl_2 as possible in both tetrahedral and square-planar forms (use an octahedral atom centre).

Why is the square-planar arrangement of atoms not compatible with experiment?*

The tetrahedral geometry for CH_2Cl_2 (Figure 1.3) therefore seems to fit the facts better than the square-planar forms (Figure 1.4). Now, there is overwhelming evidence, for example from X-ray diffraction, to support the proposition of the tetrahedral carbon atom. Every carbon atom surrounded by four atoms or groups has an approximately tetrahedral structure. No square-planar carbon compounds have ever been made (and it is unlikely that they ever will be).

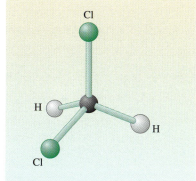

Figure 1.3
A model of tetrahedral dichloromethane (CH_2Cl_2). 🖳

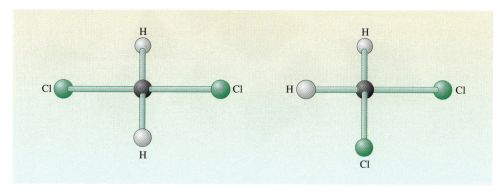

Figure 1.4 Models of the two different hypothetical square-planar forms of CH_2Cl_2.

*Answers to the model exercises start on p. 188.

BOX 1.1 van't Hoff, Le Bel and the tetrahedral carbon atom

Ideas about the stereochemistry of carbon compounds depended on the prior establishment of the valency of carbon, and on the development of structural formulae to represent molecules. J. H. van't Hoff (Figure 1.2) from the Netherlands, did his initial training at Delft Polytechnic, but he decided to abandon technology for science after a dreary vacation job in a sugar factory. J. A. Le Bel (1847–1930) was from Alsace, but he had worked in the same laboratory in Paris as van't Hoff. According to van't Hoff, it seems they 'never exchanged a word about the tetrahedron there, though perhaps both of us already cherished the idea in secret'. Le Bel was fascinated by symmetry, and developed his ideas from the writings of Louis Pasteur; van't Hoff's inspiration came from the ideas of August Kekulé, who surmised that the four valencies of carbon might be represented by four different directions (Figure 1.5a). In 1874, van't Hoff published his ideas in Dutch, but it was not until his paper was translated, first into French, then into German, that it received recognition in the chemical community. An English translation was not published until 1891.

That recognition was not all positive, however. One of the most distinguished of the older German chemists, H. Kolbe, launched a vitriolic attack on the young Dutchman. He wrote:

> One of the causes of the present regression of chemical research in Germany is the lack of general, and at the same time thorough chemical knowledge; no small number of our professors of chemistry, with great harm to the science, are labouring under this lack. A consequence of this is the spread of the weed of the apparently scholarly and clever, but actually trivial and stupid, natural philosophy, which is now brought forth again, out of the store room harbouring the errors of the human mind, by pseudoscientists who try to smuggle it, like a fashionably dressed and freshly rouged prostitute, into good society, where it does not belong.
>
> Anyone to whom this concern seems exaggerated may read, if he is able to, the book of Messrs Van't Hoff and Herrman on the *Arrangement of Atoms in Space*, which has recently appeared and which overflows with fantasies. I would ignore this book, as many others, if a reputable chemist had not taken it under his protection and warmly recommended it as an excellent accomplishment.

> H. Kolbe, *Signs of the Times* (1877)

Fortunately, this abuse did not harm van't Hoff's scientific reputation, but rather drew attention to ideas that might otherwise have been slow to spread. Nevertheless, van't Hoff did very little more work on organic stereochemistry before he turned to physical chemistry, for which he was later awarded the first Nobel Prize for Chemistry in 1901.

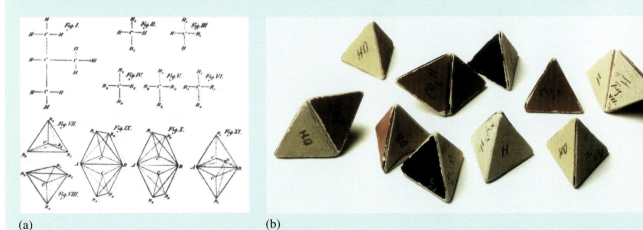

(a) (b)

Figure 1.5 (a) van't Hoff's original drawings of the tetrahedral carbon atom. Fig. II and Fig. III show the two impossible square-planar forms of a disubstituted methane; Fig. VII and Fig. VIII show the mirror image tetrahedra; and Fig. IX and Fig. X show van't Hoff's version of the *E*- and *Z*- forms of a substituted ethene, Section 5. (b) van't Hoff's space-filling molecular models.

BOX 1.2 Using molecular models

The best way of learning about molecular shape would be to observe the molecules directly. This is difficult, because of their infinitessimally small size, but the next best thing is to use molecular models. Remember in what follows that all the 'molecules' that you make are models. Molecules do not really look like coloured plastic 'blobs' held together by plastic straws! However, models of them are enormously valuable in helping us to understand certain molecular features. Most model kits are designed to show, approximately to scale, the distance apart and relative positions of the atomic centres in molecules. Usually, no attempt is made to represent the relative sizes of atoms, and the variations in the length of the bonds between different types of atoms is ignored.

Internuclear distances are very tiny: some typical interatomic distances are given in Table 1.2. Although it is difficult to comprehend, one hundred million C—H bonds laid end to end would form a line only one centimetre long. That may help to explain the difficulty we face in seeing individual molecules!

Table 1.2 Some typical covalent bond lengths; note that single-bond lengths approximate to the sum of the covalent radii in Table 1.1

Bond	Length/pm	Bond	Length/pm
H—H	74	N=O	113
C—C (ethane)	154	C—N	148
C=C (ethene)	134	C=N	127
C≡C (ethyne)	120	C≡N	117
C—H	105	N=N	124
N—H	101	C—F	138
O—H	96	C—Cl	176
C—O	143	C—Br	191
C=O	117	C—I	215
N—O	147		

Unlike those of carbon, many four-coordinate compounds of metals — particularly those of platinum, palladium and nickel — can also have square-planar geometry (in addition, nickel can form tetrahedral compounds).

MODEL EXERCISE 1.2 The molecule CXY_2Z

Use your model kit to determine how many different tetrahedral arrangements there are for CXY_2Z molecules. You will need to make at least two models to determine the answer. Save the models that you make; they will be needed again in the next Section.

An important generalization can be inferred from your answer to Model Exercise 1.2:

If two or more groups attached to a tetrahedral carbon atom are identical, then only one structure is possible.

MOLECULAR CONFORMATION

2

There is only a limited number of molecules in which a single carbon, nitrogen or oxygen centre is attached only to four, three or two atoms, respectively. One of the most significant features of the chemistry of carbon is that carbon atoms can be joined together to form long chains (a process known as *catenation*), and rings of various sizes, without the chemical compounds that are generated becoming unstable. **Straight-chain hydrocarbons** as long as $C_{78}H_{158}$ have been found in crude petroleum; poly(ethene) molecules can have carbon chains thousands of atoms long; and carbon rings of up to 30 atoms or more can be synthesised. What are the implications of the concept of the tetrahedral carbon atom when the carbon backbone is extended?

MODEL EXERCISE 2.1 Modelling the ethane molecule

Make a model of ethane, CH_3CH_3. Construct the model by joining two tetra-hedral carbon centres by a bond, and then add the other six bonds with their attached hydrogen atoms. Do not attempt to manipulate the model; just join it together and set it to one side. Next, without consulting this first model, make another ethane model in the same way. Now we can ask the question: are the two models identical?

At this stage, it might be useful to have a definition to help us decide whether two molecules are identical. For the moment we can say that two molecules are identical if they can be superimposed exactly. In other words, if a mould could be made of a molecule, an identical molecule would fit that mould in every respect. Any pair of models (or molecules!) of the type CXY_2Z in the tetrahedral form fits this criterion, whereas the two square-planar arrangements of CH_2Cl_2, that you made in Model Exercise 1.1 do not.

By handling the two ethane models and manipulating them, you probably found that you could make them identical by twisting one of the CH_3 groups while holding the other in a fixed position. A more specific description of this process is **rotation about the carbon–carbon single bond**. This **internal rotation** in the model corresponds to what happens at the molecular level.

The question that you should now be framing in your mind is 'Does this mean that there is an infinite number of different compounds called ethane, obtained by rotating one of the CH_3 groups, each with a different orientation of the hydrogen atoms attached to the two carbon atoms?'. The answer to this is definitely 'no'; there is only one compound 'ethane'. One sample of ethane always behaves in the same way as another sample of ethane, under the same conditions.

We should look at this apparent paradox a little more closely. At room temperature, rotation takes place readily about C—C single bonds (and most other single bonds) in open-chain (that is, non-cyclic) molecules; the two methyl groups in a molecule of ethane are spinning quite rapidly relative to each other. In *ring* compounds, of course, such rotation about the C—C single bonds in the ring is impossible.

STUDY NOTE

Ethane conformations on one of the CD-ROMs associated with this Book contains a video sequence showing this rotation.

The energy for this rotational motion is acquired from collisions with other molecules and with the walls of the container. If you take one of your ethane models, note its shape and then throw it at a wall; it should stay intact, but change shape slightly. Molecular collisions in real liquid and gaseous systems are very frequent, and rotations about C—C single bonds are very much easier than they appear to be in your models. Consequently, all molecules exhibit rapid *internal rotation* as they travel and tumble. Arrangements of atoms within a molecule that can be interchanged *solely by internal rotation about a single bond* are called **conformations**. *No bonds are broken or made during a change of conformation.* An individual molecule changes its conformation extremely rapidly. In this sense, there are many different possible conformations of ethane; but there is only one *compound* 'ethane', with each molecule behaving in the same way as any other.

We can now expand our definition of molecular identity.

Two, or more, molecular representations can be assumed to represent the same molecule if they can be superimposed; if necessary, *rotations about single bonds may be carried out to achieve this superimposition.*

2.1 Summary of Section 2

1   Four groups around a carbon atom take up a tetrahedral arrangement.

2   Models, and three-dimensional representations in WebLab ViewerLite, provide effective help towards understanding three-dimensional structures.

3   Identical molecules are those that can be superimposed exactly. Rotation about one or more single bonds in models may be necessary to demonstrate this.

4   Two or more arrangements that can be made identical merely by rotation about single bonds are called conformations of the same molecule.

5   The lowest-energy conformations are usually those with atoms in a staggered conformation.

6   At room temperature, rotation about a C—C single bond occurs readily.

THE REPRESENTATION OF MOLECULES

The branch of chemistry dealing with the shapes of molecules is called **stereo-chemistry**. So far, we have assumed that a molecule such as 2-methylbutane can be described unambiguously either by the molecular formula C_5H_{12}, or better, by a structural formula or a line drawing (as in Structures **3.1** and **3.2**, respectively). The molecular formula of 2-methylbutane merely tells us that the molecule consists of five carbon atoms and twelve hydrogen atoms, bonded together. Structure **3.1** gives us much more information.

The representation used in Structure **3.2** is an example of **skeletal notation**, which, unlike the structural formulae you have already met, shows no carbon or hydrogen atoms at all! However, **3.2** does show which atoms are bonded to which others. No other structural information is given by formulae or by any of these representations of molecules. These ideas are discussed further in Box 3.1.

Now that we know that a molecule is not 'flat', representations in two dimensions — in the plane of the paper — are clearly no longer entirely satisfactory. Frequently, we need more than this; specifically, we need a representation that conveys a molecule's shape and perspective more accurately. A compound written as XYZCCABD (see Structure **3.3**) can indeed have many different conformations; however, it could also be one of several different structures, as you will see.

This is why we need to sort out the problem of representing three-dimensional structures on two-dimensional paper. In general, the most useful of the methods available is the picturesquely named **'flying-wedge' notation**. This uses three different types of line to represent bond direction — the wedge, the continuous line, and the broken line. A flying-wedge representation of dichloromethane is shown in Figure 3.1 (the central 'C' is sometimes omitted).

The carbon atom at the pointed or thin end of the wedge (——◣) is assumed to be in the plane of the paper, and the atom at the thick end is *above the plane of the paper*. The connection between the shape of this wedge and the observed perspective is obvious. A continuous line (—) joins two atoms that both lie in the plane of the paper; in Figure 3.1 there are two such lines, between the carbon atom and the two hydrogen atoms. A broken line (– –) joins together two atoms, one of which is in the plane of the paper (carbon) and the other (chlorine) below it. Any atom at the junction of a broken line with a wedge, or at the junction of two continuous lines, will therefore lie in the plane of the paper. The orientation in space of the dichloromethane molecule in Figure 3.1 is exactly the same as the orientation of the molecular model of methane shown in Figure 1.1.

● Draw another flying-wedge representation of dichloromethane, but this time with the two chlorine atoms and the carbon atom in the plane of the paper.

● There are many possibilities, two of which are shown as Structures **3.4** and **3.5**.

3.1

$$H_3C - \underset{\underset{H}{|}}{\overset{\overset{CH_3}{|}}{C}} - CH_2CH_3$$

3.2

3.3

$$Y - \underset{\underset{Z}{|}}{\overset{\overset{X}{|}}{C}} - \underset{\underset{D}{|}}{\overset{\overset{A}{|}}{C}} - B$$

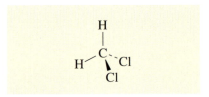

Figure 3.1
A flying-wedge representation of dichloromethane.

3.4

3.5

BOX 3.1
The representation of molecules by abbreviated structural formulae and line drawings

Organic chemists do not often draw out representations of molecules showing all the bonds; instead, they tend to use condensed or abbreviated forms. The abbreviated structural formula of the C_5H_{12} molecule shown in Structures **3.1** and **3.2** would usually be written $(CH_3)_2CHCH_2CH_3$, and the straight-chain molecule hexane, C_6H_{14}, would be written $CH_3(CH_2)_4CH_3$; it is represented in skeletal notation in Structure **3.6**. You have encountered this approach already with functional groups, where $-NO_2$ is written instead of

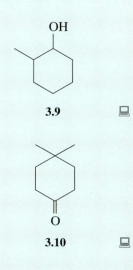

3.6

and $-COOH$ (or CO_2H) is written instead of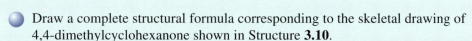

For cyclic compounds, it is less easy to abbreviate formulae in this way. Cyclohexane (Structure **3.7**) must still be written out in full (rather than as $(CH_2)_6$, say), so another convention is used. The six-membered ring of CH_2 groups is drawn as a hexagon (Structure **3.8**); this is another example of skeletal notation.

3.7   **3.8**

In skeletal representations such as Structure **3.8**, you have to imagine that each line represents a bond — in particular, a carbon–carbon bond — unless another atom is inserted. You have to remember that two hydrogens must be mentally attached at every angle in the drawing; if three lines meet at a point, then only one hydrogen must be attached. A line at the end of any structure terminates in a methyl group, CH_3. Functional groups are then attached to this carbon skeleton. So 2-methylcyclohexanol is represented as in Structure **3.9**.

OH

3.9

● Draw a complete structural formula corresponding to the skeletal drawing of 4,4-dimethylcyclohexanone shown in Structure **3.10**.

● As indicated in Structure **3.11**, a complete structural formula includes all the atoms in the molecule.

O

3.10

3.11

It should be stressed that the representations of dichloromethane shown in Structures **3.4** and **3.5**, and the one shown in Figure 3.1, all show exactly the same molecule with exactly the same bond lengths and bond angles; they simply represent three different views of the same molecule.

MODEL EXERCISE 3.1 The molecule CH₂BrCl

Make a model of bromochloromethane, CH_2BrCl, and use this to help you to draw flying-wedge representations of the molecule from a number of different viewpoints. You will probably be surprised at the number of different representations that can be made of the one molecule!

Of course, it is not necessary to draw 18 (or more) representations of the one molecule. Each figure contains exactly the same information, so one figure is sufficient to define the molecule. It is important to be able to draw the figures from a number of views, however, because with more complex molecules one needs to be able to choose a view that conveys the structural information with the greatest clarity.

🔵 Now try to draw a flying-wedge representation of the two conformations of the ethane molecule shown in Figure 3.2.

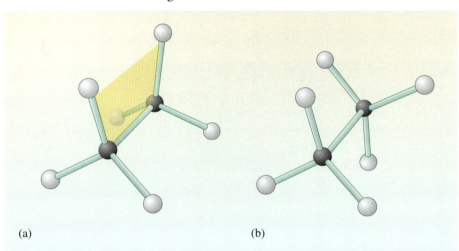

(a)   (b)

Figure 3.2
The two models of ethane used in Model Exercise 2.1 (in model (a), corresponding C—H bonds attached to the front and back carbon atoms are in the same plane, as exemplified by the yellow-shaded area).

⚪ The closest representations are shown in Structures **3.12** and **3.13**.

3.12   **3.13**

The form shown in Structure **3.14** is sometimes described as the 'sawhorse' conformation (although it has only a passing resemblance to this carpenter's tool, even when drawn without atom symbols, as in Structure **3.15**). The different representations **3.12**–**3.15** show clearly that even in two dimensions we can readily distinguish different conformations. With a little practice you will find that you can easily draw, and interpret, these flying-wedge representations. You will also discover that they are a great help in understanding organic chemistry in three dimensions.

3.14   **3.15**

COMPUTER ACTIVITY 3.1
Drawing and manipulating computer models

In this Computer Activity you will use ISIS/Draw and WebLab ViewerLite to reproduce the two different conformations corresponding to the models of ethane shown in Figure 3.2, and Structures **3.12** and **3.13**.

The Activity in *Exploring the third dimension* (on one of the CD-ROMs) should take you approximately 20 minutes to complete.

Another way of representing conformations is the **Newman projection**. These have a more restricted use, but they are an invaluable complement to the flying-wedge notation. A Newman projection shows a view of a molecule looking along the line of one bond *from the left*, for example the C—C bond in ethane. The nearer carbon atom of that bond is represented by a point, and the bonds to the three groups attached to it by straight lines at 120° to each other (Structure **3.16**). The further atom of the C—C bond is shown as a circle, with the bonds to its three attached groups as lines to the circumference (Structure **3.17**). The two are then superimposed to give the Newman projection.

Newman projections and flying-wedge representations of the two conformations of ethane in Figure 3.2 are shown in Figure 3.3.

nearer carbon atom
in Newman projection

3.16

further carbon atom
in Newman projection

3.17

Figure 3.3 Flying-wedge representations, and the corresponding Newman projections, of the ethane molecule in the conformations shown in Figure 3.2a and b.

It is not so easy to draw the Newman projection in Figure 3.3a, in which the hydrogen atoms of one methyl group are 'hiding behind' the hydrogen atoms of the other methyl group, as viewed along the C—C bond (see also Figure 3.2a). This is called the **eclipsed conformation**. The conformation that shows the hydrogen atoms attached to the further carbon atom positioned exactly between the hydrogen atoms attached to the nearer carbon atom (as in Figures 3.2b and 3.3b), is called the **staggered conformation**.

QUESTION 3.1

Which of the following Newman projections (a)–(c) represents the identical conformation to the flying-wedge representation of 1,2-dichloroethane in Structure **3.18**?

3.18 1,2-dichloroethane

(a)

(b)

(c)

QUESTION 3.2

Draw Newman projections and flying-wedge representations of the chloroethane and 1,2-dichloroethane models shown in Figure 3.4.

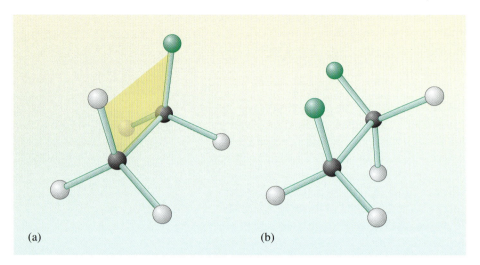

(a)   (b)

Figure 3.4 Models of (a) chloroethane, and (b) 1,2-dichloroethane.

It is important for you to be able to represent molecules using both these conventions, and to be able to convert flying-wedge representations into Newman projections, and vice versa.

MODEL EXERCISE 3.2 Resistance to rotation

This exercise shows a limitation of model kits. Take one of your models of ethane, and rotate one methyl group by 360° relative to the other. Can you detect any change in resistance to rotation as you rotate the model through different conformations?

Although models cannot show it, some molecular conformations have more internal energy than others. The staggered conformation in ethane (Figures 3.2b and 3.3b), and almost all other molecules, is the one that has the lowest energy, and so is favoured. This is because atoms have a finite size (see Table 1.1), and non-bonded atoms tend to dispose themselves as far away from each other as possible, within the constraints of fixed bond lengths. This is analogous to the clustering of atoms or groups about a single centre, as discussed earlier. In molecules with more than one centre, there is also repulsion between chemical bonds, because of the bonding electrons. This has the same effect of favouring the staggered conformation.

The difference in energy between staggered and eclipsed forms in ethane is very small, but even so it is sufficient to ensure that an ethane molecule spends nearly all its time in the staggered conformation. As the size of substituents increases, so does the energy difference between the eclipsed and staggered conformations. Nevertheless, at normal temperatures, individual conformations of most substituted ethanes cannot be isolated, since bond rotation is so fast.

When atoms or groups other than hydrogen are attached to a carbon–carbon bond, a wider range of more-stable conformations is possible. If the temperature of a sample is lowered, so is its overall kinetic energy. As conformational rotation slows down, it may be possible to observe different low-energy conformations using spectroscopic methods. You will be able to explore the range of possible conformations in Computer Activity 3.2.

COMPUTER ACTIVITY 3.2
Newman projections in ISIS/Draw

This is an Activity which introduces you to more of the functions of ISIS/Draw. In this Computer Activity you will practise drawing Newman projections using ISIS/Draw, and will be introduced to the whole range of possible conformations.

The Activity (in *Exploring the third dimension* on one of the CD-ROMs) should take you approximately 25 minutes to complete.

As you can see in Figure 3.5, some other specific conformations are given particular names.

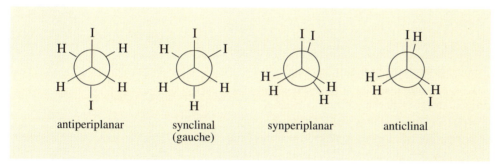

antiperiplanar   synclinal (gauche)   synperiplanar   anticlinal

Figure 3.5 Newman projections of different named conformations in 1,2-diiodoethane. 💻

There are two different staggered conformations. The one in which the two iodine atoms are displaced at 180° to one another in the Newman projection is said to have the iodine atoms **antiperiplanar**. (You may also see the word *trans* used in some texts.) The other staggered conformation in which the iodine atoms are 60° apart is called the **synclinal conformation** (also called *gauche*). The iodine atoms in the antiperiplanar conformation are further apart than they are in the synclinal conformation.

The two eclipsed conformations also have special names. When the iodine atoms overlap, the conformation is called **synperiplanar**. The conformation in which hydrogen and iodine atoms overlap is called **anticlinal**. The terms, syn- and anti-periplanar and syn- and anticlinal are general terms that relate to any compound.

QUESTION 3.3

Draw Newman and flying-wedge projections of: (a) a synclinal conformation of l-bromo-2-chloroethane, CH_2BrCH_2Cl; (b) an antiperiplanar conformation of butane, $CH_3CH_2CH_2CH_3$; (c) an eclipsed conformation of propane, $CH_3CH_2CH_3$; (d) an anticlinal conformation of 1,2-dibromoethane, CH_2BrCH_2Br; (e) a synperiplanar conformation of 2-bromoethan-1-ol, CH_2BrCH_2OH.

Conformations having the lowest internal energies are called **preferred conformations**. If you tried to identify which of the two conformations in Figure 3.2 has the higher internal energy, you may not have found it easy to decide. The following Computer Activity with ISIS/Draw and WebLab ViewerLite should help you to reach a decision.

COMPUTER ACTIVITY 3.3
The stereochemistry of 1,2-diiodoethane

In this Activity (in *Exploring the third dimension* on one of the CD-ROMs) you will look at the conformations of 1,2-diiodoethane, and measure the non-bonded distances between atoms in staggered and eclipsed conformations. Comparing these distances with the van der Waals radii of the hydrogen and iodine atoms should help you to understand the concept of the preferred conformation for a molecule.

Question 3.4 also relates to this Activity, which should take you about 25 minutes to complete.

QUESTION 3.4

Compare the sum of two van der Waals radii for iodine atoms with the iodine–iodine non-bonded interatomic distance in both the synclinal and the synperiplanar (iodine–iodine eclipsed) conformations; and the sum of the van der Waals radii of iodine and hydrogen atoms with the iodine–hydrogen distance in the antiperiplanar conformation and the anticlinal (iodine–hydrogen eclipsed) conformation. Use these interatomic distances to decide which of the four conformations will have the highest internal energy, and explain why. Enter your values in the following table.

Conformation	Iodine atoms antiperiplanar	Iodine atoms synperiplanar	Iodine atoms synclinal	Iodine eclipsed with hydrogen
I–I distance/pm	not needed			
I–H distance/pm				
H–H distance/pm				

sum of van der Waals radii: I + I/pm

I + H/pm

H + H/pm

As you should have discovered in Question 3.4, the synperiplanar conformation, in which the two large iodine atoms are eclipsed, has the shortest iodine–iodine interatomic distance, and so it will have the highest energy. As the iodine atoms move apart through rotation about the carbon–carbon bond, the iodine–iodine interatomic distance increases. Even in the synclinal conformation, the two iodine atoms still interact quite strongly, but in the antiperiplanar conformation they are at their maximum distance apart. In this conformation the iodine–hydrogen interaction is minimized too, so the antiperiplanar conformation will have the lowest internal energy.

From this analysis we can formulate two general rules for identifying preferred conformations:

1   If only two large atoms or groups are attached at different ends of a carbon–carbon bond (the other groups being H), the preferred conformation will usually be that in which the two large groups are as far apart as possible — that is, the antiperiplanar conformation.

2   If groups of different sizes (small (S), medium (M) and large (L)) are attached to both ends of a carbon–carbon bond, the preferred conformation will usually be the one in which the large group at one end of the bond lies between the small and medium groups at the other end of the bond, in a staggered conformation (see, for example, Structure **3.19**).

3.19
preferred conformation

To apply these rules we also need to know something about the sizes of some of the common organic groups. These sizes are not usually expressed in picometres like van der Waals radii, but a generally accepted order of size is as follows:

$$(CH_3)_3C > C_6H_5 > I > Br >> CH_3 > Cl > NO_2 > COOH >> OCH_3 > OH > F > H$$

This phenomenon, where the shape of a molecule — or sometimes its reactivity — is limited by the interaction of electron clouds on non-bonded atoms or groups, is known as **steric hindrance**.

QUESTION 3.5

Using data from Table 1.1, draw Newman projections of the preferred conformations of the following molecules: (a) 1-chloro-2-phenylethane, $C_6H_5CH_2CH_2Cl$ (Structure **3.20**); (b) 2-chloro-3-phenylbutane, $CH_3CH(Cl)CH(C_6H_5)CH_3$ (Structure **3.21**).

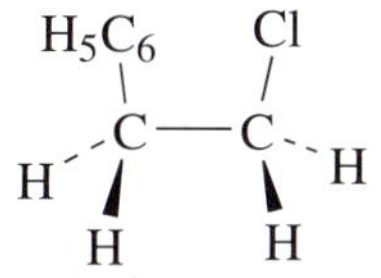

3.20

MODEL EXERCISE 3.3   Modelling poly(ethene)

Polythene, or poly(ethene), is a polymer of ethene, which has the general formula $(CH_2\!-\!)_n$, where n can be of the order of 10^5. Use your model kit to construct a short polythene chain with ten carbon atoms in it, equivalent to $-CH_2(CH_2)_8CH_2-$. Make a model of this chain in the *lowest-energy conformation possible*. What do you think this implies about low-energy conformations for the carbon backbone of long-chain molecules, and for the behaviour of polymer molecules themselves?

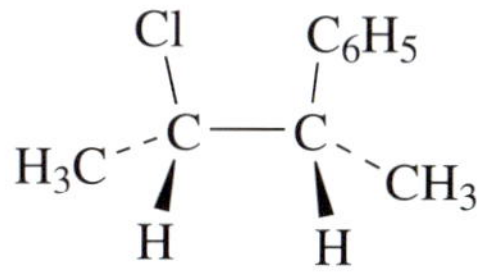

3.21

CONSTITUTIONAL (STRUCTURAL) ISOMERISM

4

Before we go on to study some further consequences of the three-dimensional structures of molecules, we must pause to consider again the ambiguity that exists in trying to describe a molecule only by its molecular formula. We have seen that there can be only one molecule with the formula CH_2Cl_2, and only one with the formula CH_3CH_3, but what about molecules that have greater numbers of carbon atoms in their structure?

● What is the smallest number of carbon atoms for which more than one carbon skeleton can be constructed, while retaining the normal requirement of each carbon atom having four bonds attached to it?

● There is only one way that three carbon atoms can be connected together in a chain (as propane, **4.1**), but three carbon atoms can also form a ring (cyclopropane, **4.2**).

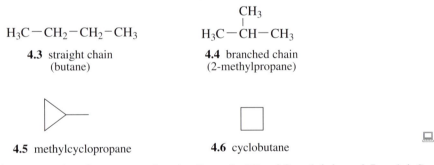

$CH_3CH_2CH_3$

4.1 propane   **4.2** cyclopropane 🖳

With four carbon atoms, however, there are two distinct possible non-cyclic arrangements, the straight chain (**4.3**), and the **branched chain** (**4.4**) as well as two cyclic possibilities, methylcyclopropane (**4.5**) and cyclobutane (**4.6**).

$$H_3C-CH_2-CH_2-CH_3$$

4.3 straight chain
(butane)

$$\overset{\overset{\displaystyle CH_3}{|}}{H_3C-CH-CH_3}$$

4.4 branched chain
(2-methylpropane)

4.5 methylcyclopropane

4.6 cyclobutane

🖳

Molecules possessing the same molecular formula (like **4.3** and **4.4**, or **4.5** and **4.6**), but having different structures are called **isomers**. Isomers are distinct compounds, often with very different physical and chemical properties. Unlike *conformations*, which can be interconverted simply by rotation about internal bonds, one isomer can only be converted into another by breaking and remaking covalent bonds to give a different arrangement of the atoms. Alternative structures in which only the carbon skeleton is changed are called **skeletal isomers**.

QUESTION 4.1

How many different carbon skeletons can be built from *six* carbon atoms? Limit yourself to skeletons with carbon chains and *no more than one ring*. Will all your skeletons require the same number of hydrogen atoms if they are *saturated* hydrocarbons? You will need to jot down your structures on a sheet of paper.

If atoms other than carbon and hydrogen are included in the formula, a different form of isomerism is possible.

⬤ What other dichloroethane is possible besides the 1,2-dichloroethane (**3.18**) discussed in Question 3.1?

⬤ The isomer 1,1-dichloroethane (Newman projection (b) in Question 3.1) is also possible.

Isomers like this, in which the atoms or functional groups are placed at different sites within the same carbon skeleton, are called **positional isomers**. So, for example, Structures **4.7** and **4.8** are positional isomers, because they have the same carbon backbone but the chloro- functional group is attached at different positions.

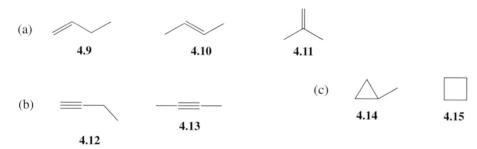

4.7   **4.8**

It is, of course, possible to have both skeletal and positional isomers from one molecular formula, as Question 4.2 reveals.

QUESTION 4.2

How many isomers, skeletal and positional, are possible for the formula C_4H_8ClBr? Again, pen and paper are essential to answer this question.

Positional isomerism is also possible in unsaturated molecules, where the functional group that can occur in different positions within the carbon skeleton is a double or triple bond. Once again, the way to ensure you have all the isomers is to draw every possible carbon skeleton, and then place the double or triple bond in every possible position.

⬤ How many skeletal structures are possible for a C_4 molecule with either (a) one double bond, (b) one triple bond, or (c) a single cyclic system?

⬤ The seven possible isomers are Structures **4.9**–**4.15**.

(a)

4.9   **4.10**   **4.11**

(b)

4.12   **4.13**

(c)

4.14   **4.15**

⬤ From Structures **4.9**–**4.15**, work out the molecular formulae of (a) the open-chain compounds that include a double bond; and (b) the cyclic structures. How does each differ from the formula of butane, the saturated C_4 hydrocarbon?

⬤ Both the unsaturated hydrocarbons and the cyclic hydrocarbons have the formula C_4H_8 — that is, two hydrogens less than the saturated molecule butane, C_4H_{10}.

This is a general rule; compounds that include either one double bond, or one ring system, in their structure, always have two fewer hydrogen atoms in their molecular formula than the equivalent saturated compound. This feature is cumulative, so a compound with two double bonds or two rings (or one double bond and one ring) has four fewer hydrogen atoms; a compound with three double bonds (or one double and one triple bond, or two double bonds and one ring) has six fewer. Nor does it matter which atoms are linked by the double bond. The introduction of a C=N or C=O bond instead of C−N or C−O also reduces the hydrogen count by two.

There is a third type of isomerism, **functional isomerism**, which is possible when atoms, like oxygen, nitrogen and sulfur, which can occur in more than one type of functional group are included in the formula. The simplest example of these is a formula with one oxygen atom, like C_2H_6O. Using the abbreviated formula, two isomers can be identified — CH_3CH_2OH, with an alcohol functional group, and CH_3OCH_3, with an ether functional group[*]. If the formula indicates unsaturation in the molecule, remember that double bonds can occur in either the functional group(s) or in the carbon skeleton; cyclic structures are also possible.

○ Which three structures can be written for the formula C_2H_4O?

○ They are ethanal, **4.16**, ethenol, **4.17** (sometimes given the name vinyl alcohol), and the cyclic ether, **oxirane**, **4.18** (or ethylene oxide, as it is known in the chemical industry):

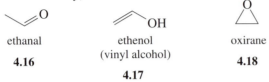

ethanal

4.16

ethenol
(vinyl alcohol)

4.17

oxirane

4.18

We have described the three forms of isomerism — skeletal, positional and functional — which collectively are described as **constitutional isomerism** (sometimes called **structural isomerism**). Constitutional isomers are molecules that have the same molecular formula, but with the atoms connected in a different sequence. Each constitutional isomer will therefore have a distinct chemical name.

QUESTION 4.3

How many different alcohols and how many different ethers can be drawn for the formula $C_4H_{10}O$? Remember to start by drawing all the possible skeletons for a structure with four carbon atoms and one oxygen atom. Give your structures systematic names to ensure that they are all different.

QUESTION 4.4

Draw all the isomers — skeletal, positional and functional — that have the molecular formula $C_3H_6O_2$. Don't forget that unsaturation can occur in functional groups as well as in the carbon skeleton. Name the functional groups in each isomer.

EXERCISE 4.1
Summarizing constitutional (structural) isomerism

Try to express the different forms of constitutional (structural) isomerism in the form of a diagram, showing both the types of isomers and their characteristics.

---

[*] Functional groups are discussed in the Appendix, p. 169.

STEREOISOMERS OF MOLECULES CONTAINING DOUBLE BONDS

5

We have seen in the preceding Section that compounds containing double bonds can exhibit positional isomerism; for example, but-1-ene (**4.9**) and but-2-ene (**4.10**) were two such positional isomers in the earlier discussion. However, we have implicitly, *and incorrectly*, assumed that by writing the structural formula $CH_3CH\!=\!CHCH_3$ for but-2-ene, its structure has been uniquely defined. In fact, there are *two* quite distinct compounds to which the general formula $R^1CH\!=\!CHR^2$ is applicable. This arises from the rigid nature of the carbon–carbon double bond. Unlike the carbon–carbon single bond, where rotation is easy, rotation about double bonds is not possible under normal conditions. This phenomenon is referred to as **restricted rotation**. At this stage you needn't be concerned about the origin of the rigidity. You may simply assume that any fragment $XZC\!=\!CYW$ (Structure **5.1**) is rigid, and that all six atoms lie in a plane. All the angles between the bonds in the structural formula are about 120°.

X 120° Y
    120°
Z   W

5.1

MODEL EXERCISE 5.1
Modelling molecules with double bonds

You can simulate this rigidity when you make models of alkenes with your model kit. The construction of a carbon–carbon double bond is shown in Figure 5.1. In model kits such as the Orbit system, rotation about the $C\!=\!C$ double bond is prevented by attaching white 'pegs' to a second green straw, and locking the two carbon atoms together by inserting the pegs into the hole in the middle of both of the three-coordinate centres (Figure 5.1a).

Alternatively, take two tetrahedral carbon centres, and join two prongs on one to two prongs on the other, using two flexible white straws (Figure 5.1b). All model kits have some device for preventing rotation about double bonds in models of molecules which include them.

But-2-ene, $CH_3CH\!=\!CHCH_3$, has two isomers; make models of them, and examine them closely. How are the two methyl groups arranged about the double bond in each isomer?

Now make models of the isomers of 2-methylbut-2-ene, $(CH_3)_2C\!=\!CHCH_3$. How many different models can you make?

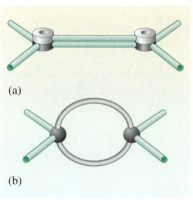

(a)

(b)

Figure 5.1
How to make a $C\!=\!C$ double bond with the Orbit model kit: (a) using pegs and a second green straw between the carbon centres; (b) using two flexible white straws for the double bond. Other model kits may use different procedures.

Compounds **5.2** and **5.3** are not able to interconvert rapidly, even at temperatures greater than 200 °C. They are therefore not different conformations of the same molecule. Instead, they are described as the two **configurations** of the molecule $CH_3CH\!=\!CHCH_3$. As with all isomers, one configuration can only be converted into another configuration by the making and breaking of bonds. That is what distinguishes conformations from configurations.

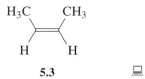

H_3C H
H CH_3

5.2

H_3C CH_3
H H

5.3

Such pairs of molecules are members of a class of isomers called **stereoisomers**, which are defined as distinct, separable compounds with identical constitutions (or connectivity), but with different spatial arrangements of their atoms. Although the atoms of these compounds are joined together in the same sequence from one end of the molecule to the other, they clearly have different three-dimensional structures. In addition, the name **geometrical isomers** is used for stereoisomers like the two but-2-enes (**5.2** and **5.3**), which differ only in the spatial orientation of their constituent groups.

The observations in Model Exercise 5.1 allow us to define a criterion for geometrical isomerism about a carbon–carbon double bond.

- Before reading further, see if you can formulate your own rule to decide, from its skeletal or abbreviated structural formula, whether a molecule with a carbon–carbon double bond is capable of exhibiting geometrical isomerism or not.

- Geometrical isomerism is only possible if there are two different groups attached to the left-hand end of the C=C double bond, and two different groups attached to the right-hand end of the double bond.

In general terms, molecules having one of the formulae WXC=CYZ, WXC=CXZ, or WXC=CWX are able to exhibit geometrical isomerism. There is, however, only one isomer for molecules with either of the formulae WXC=CW$_2$ or W$_2$C=CW$_2$.

QUESTION 5.1

Use drawings to show which of the following molecules may exist as geometrical isomers: 3-phenylpropenoic acid (cinnamic acid) $C_6H_5CH=CHCOOH$; 1,2-diphenylethene (stilbene) $C_6H_5CH=CHC_6H_5$; hex-1-ene; hex-2-ene.

Question 5.1 shows that we are not able to use an abbreviated structural formula to describe unambiguously a molecule that can exhibit geometrical isomerism. So we need a more accurate means of specifying the geometry, or configuration, of a particular alkene. There are two methods in common use.

For simple alkenes HXC=CYH (where X and Y are not H) in which the two hydrogen atoms are on the *same* side of the double bond, the label *cis* is applied; if the two hydrogen atoms are on *opposite* sides of the double bond, the designation is *trans* (cf. *trans*atlantic — meaning on opposite sides of the Atlantic Ocean). The two 1,2-diphenylethene structures shown in the answer to Question 5.1, are re-drawn below as Structures **5.4** and **5.5** with their *cis* and *trans* designations.

5.4 *trans*-1,2-diphenylethene   **5.5** *cis*-1,2-diphenylethene

Strictly speaking, having *hydrogen* atoms on both double bond carbon atoms is not necessary. As long as there is the *same atom or group* on both ends of the C=C bond, the *cis/trans* designation can be used; for example, 2,3-dichlorobut-2-ene, $CH_3CCl=CClCH_3$, has *cis* and *trans* isomers (Structures **5.6** and **5.7**), as does 1,2-dichloro-1-phenylprop-1-ene, $C_6H_5CCl=CClCH_3$ (Structures **5.8** and **5.9**).

5.6 *cis*-1,2-dichlorobut-2-ene

5.7 *trans*-1,2-dichlorobut-2-ene

5.8 *cis*-1,2-dichloro-1-phenylprop-1-ene

5.9 *trans*-1,2-dichloro-1-phenylprop-1-ene

Draw and label the structures for *cis*- and *trans*-1,2-diiodopropene.

These compounds are shown as Structures **5.10** and **5.11**.

5.10 *trans*-1,2-diiodopropene

5.11 *cis*-1,2-diiodopropene

The preferred and general system for describing alkene configurations is a special nomenclature system called the **Cahn–Ingold–Prelog system**, after its inventors, R. S. Cahn, C. K. Ingold and V. Prelog. In this system (first proposed in 1956), the two groups attached to the carbon atoms at each end of the carbon–carbon double bond are each given a priority.

The two main **priority rules** for determining the priority of attached atoms are:

(a) An attached atom of higher atomic number has priority over one of lower atomic number (for example, Cl > F > O > N > C).

(b) Where the two atoms immediately bonded to a carbon atom of the double bond are identical, further rules must be applied. For example, if both directly attached atoms are carbon, a carbon atom bonded to halogen, oxygen, nitrogen or sulfur has priority over a carbon atom bonded to another carbon; so —CH$_2$OH has priority over —CH$_2$CH$_3$. A larger alkyl group has priority over a smaller alkyl group, so ethyl has priority over methyl; and the carbon atom of a benzene ring has priority over a double-bonded carbon atom of an alkene, which itself has priority over a singly bonded carbon. An atom attached through a double bond is counted twice, and that through a triple bond is counted three times, so attached —C=C counts as two carbon atoms, and —C≡N as three nitrogen atoms, for example.

These rules generate the order of priorities given in Figure 5.2, which also appears in the *Data Book* (from one of the CD-ROMs associated with this Book). Using this Figure, you should be able to overcome any problems of priority allocation.

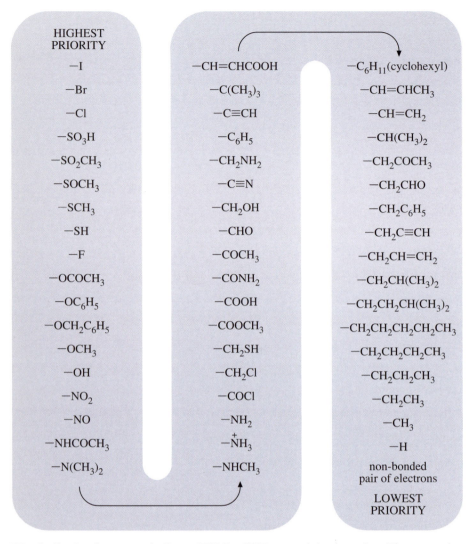

HIGHEST PRIORITY

—I

—Br

—Cl

—SO₃H

—SO₂CH₃

—SOCH₃

—SCH₃

—SH

—F

—OCOCH₃

—OC₆H₅

—OCH₂C₆H₅

—OCH₃

—OH

—NO₂

—NO

—NHCOCH₃

—N(CH₃)₂

—CH=CHCOOH

—C(CH₃)₃

—C≡CH

—C₆H₅

—CH₂NH₂

—C≡N

—CH₂OH

—CHO

—COCH₃

—CONH₂

—COOH

—COOCH₃

—CH₂SH

—CH₂Cl

—COCl

—NH₂

—N⁺H₃

—NHCH₃

—C₆H₁₁ (cyclohexyl)

—CH=CHCH₃

—CH=CH₂

—CH(CH₃)₂

—CH₂COCH₃

—CH₂CHO

—CH₂C₆H₅

—CH₂C≡CH

—CH₂CH=CH₂

—CH₂CH(CH₃)₂

—CH₂CH₂CH(CH₃)₂

—CH₂CH₂CH₂CH₂CH₃

—CH₂CH₂CH₂CH₃

—CH₂CH₂CH₃

—CH₂CH₃

—CH₃

—H

non-bonded pair of electrons

LOWEST PRIORITY

Figure 5.2
Order of priority of atoms and groups for application of the sequence rules.

We shall take the general alkene WXC=CYZ to explain the rules. There are three steps:

(a) First, determine the priorities of W and X, the two substituent atoms or groups attached to one of the double bond carbon atoms. Label them 1 and 2, where 1 has higher priority than 2 (as shown in Structure **5.12**).

(b) Then, do the same for Y and Z, the substituents on the other carbon atom of the double bond (as, for example, in Structure 5.12) .

(c) Finally, compare the positions of the two higher-priority groups — that is, those groups with priority 1.

Where the two groups with priority 1 are on *opposite* sides of the double bond, the configuration is given the label **E** (from the German word *entgegen* meaning 'opposite'). When the two higher-priority groups are on the *same* side of the double bond (as in Structure 5.12), the label **Z** (from the German word *zusammen*, meaning 'together') is applied.

Let's try this for 3-phenylpropenoic acid (cinnamic acid). At the left-hand end of the double bond, the phenyl group, C₆H₅ (1), has priority over hydrogen (2); and at the right-hand end, carboxylic acid (1) has priority over hydrogen (2). So Structure **5.13**

5.12

is given the designation *E*-, because the groups of higher priority are on opposite sides of the double bond; Structure **5.14** is designated *Z*-.

5.13 *E*-cinnamic acid

5.14 *Z*-cinnamic acid

Now for a more difficult example.

⬤ What is the configuration of the alkene in Structure **5.15**?

⬤ Looking first at the left-hand carbon atom in the diagram, Br (1) has higher priority than CH_3 (2). On the right-hand carbon atom, C_6H_5 (1) has higher priority than CH_3 (2). The two higher-priority groups, Br and C_6H_5, are on the same side of the double bond, so the alkene is designated *Z*. Its full name would therefore be *Z*-2-bromo-3-phenylbut-2-ene.

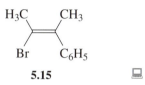

5.15

QUESTION 5.2

Draw the possible isomers of the following compounds (i)–(v), labelling them as *E* or *Z* (and *cis* or *trans* if appropriate):

(i) $ClHC=CHCl$; (ii) $Cl(CH_3)C=CH_2$; (iii) $(CH_3)_2C=CBrCl$;
(iv) $(C_6H_5)HC=CHCH_3$; (v) $(C_6H_5)ClC=CBrCH_3$.

Before an alkene can exist as *cis/trans* (or *E/Z*) isomers, each carbon atom of the double bond must bear two different groups. It does not matter if the groups on both the carbon atoms are the same, as in (i) in Question 5.2, as long as the individual carbon atoms each bear two different groups.

Surprisingly, geometrical isomerism has significance in the insect world; see Box 5.1.

One interesting consequence of restricted rotation about carbon–carbon double bonds occurs in the geometry of ring systems containing them. The configuration of the double bond in 4-, 5-, and 6-membered rings is necessarily *Z* (or *cis*); see Structures **5.16**–**5.18**. It is impossible to make a reasonable model of an *E* (or *trans*)-configuration of any of these rings, just as it has proved impossible to synthesise the compounds *E*-cyclobutene, *E*-cyclopentene, or *E*-cyclohexene; the cycloalkene rings are just too strained.

5.16 cyclobutene   **5.17** cyclopentene

5.18 cyclohexene

MODEL EXERCISE 5.2
Trans double bonds in cycloalkenes

Use your model kit to find how many carbons are needed in a ring before a reasonably strain-free model of a *trans*-cycloalkene can be made. Include the hydrogen atoms in your model, since these help to illustrate the severe crowding. This exercise provides the first example of how you can use molecular models to predict chemical behaviour.

BOX 5.1 Pheromones

Many insects communicate with others of their species by releasing minute amounts of specific chemical compounds called **pheromones**. Among others, there are alarm pheromones that signal danger, sex-attractant pheromones that enable males and females to locate each other for mating, and trail pheromones that signal the location of food.

The receptors that detect the sex-attractant pheromones are extremely sensitive. Reception of just a few hundred molecules released by a female may be enough to stimulate the male! Experiments have shown that in some species the male insect can be lured from distances as great as four kilometres, and that a single female pine saw-fly attracted more than 11 000 males! Moreover, recent research has shown that the same pheromone may be used by more than one species; for example, the cabbage looper moth and the Asian elephant share a common sex attractant! The diamondback moth (Figure 5.3) is unusual in that the female of this species produces two sex attractants. The structures of these and some other pheromones are shown in Table 5.1.

Figure 5.3
The diamondback moth.

Table 5.1 Insect pheromones

$CH_3CH_2CH_2CH{=}CHCHO$ (hex-2-enal)	ant alarm
$(CH_3)_2C{=}CHCH_2CH_2C(CH_3){=}CHCH_2CH_2C({=}CH_2)CH{=}CH_2$ (β-farnesene)	aphid alarm
$CH_3(CH_2)_3CH{=}CH(CH_2)_6OCOCH_3$	cabbage looper moth sex attractant
$CH_3CH_2CH_2CH{=}CHCH{=}CH(CH_2)_8CH_2OH$	silkworm moth sex attractant
$CH_3(CH_2)_4CH{=}CH(CH_2)_4CH_2OCOCH_3$	sugar-beet moth sex attractant
$CH_3(CH_2)_3CH{=}CH(CH_2)_9CHO$ and $CH_3(CH_2)_3CH{=}CH(CH_2)_9CH_2OCOCH_3$	diamondback moth sex attractants

A common feature of many of these compounds is the presence of one or more carbon–carbon double bonds, and the configuration of these bonds, *E*- or *Z*-, is critical for the pheromone's function. What's more, it is sometimes found that the presence of other isomers can even cancel out the effect of a pheromone. The ant alarm hex-2-enal is only active in the *trans* configuration, for instance, and the female sugar-beet moth sex pheromone has to be in the *cis* configuration to attract the males; both the diamondback moth sex attractants have the *cis* configuration.

Aphids (like the common greenfly and blackfly) are the major insect pest of arable farming in Britain, causing loss and damage by their feeding, and also as transmitters of plant diseases. They are usually controlled by 'contact' insecticides, but as they tend to cluster on the underside of leaves, they can avoid coming into contact with sprayed insecticide unless high concentrations are used. However, if an alarm pheromone is included in the spray, this would make the aphids more active, and more likely to encounter the insecticide (Figure 5.4).

syringe
needle

Figure 5.4 Aphids (left) are rapidly dispersed by the application of one drop of alarm pheromone (right).

Since only minute amounts of the natural material is available, the first challenge for chemists is to identify the pheromone, and then to synthesise it. This problem has been tackled at the Rothampsted Experimental Station in Hertfordshire, where chemists identified the aphid alarm pheromone β-farnesene by separating the various components of secretions from disturbed aphids. The true alarm pheromone caused a nerve impulse to activate an electrode attached to the sensory organ of a live, captive aphid.

The structure of β-farnesene contains four carbon–carbon double bonds, only one of which can adopt a *cis* or *trans* configuration. Both isomers were synthesised, but only the *trans* compound evoked a response; the *cis* compound was completely inactive as a pheromone. Both configurations are shown in the WebLab ViewerLite image in Figure 5.5. Their overall shapes are quite different. As the functional groups are the same for both isomers, we may assume that it is the shape of the insect receptor which determines the pheromone activity. Only the *trans*-β-farnesene isomer must fit the receptor!

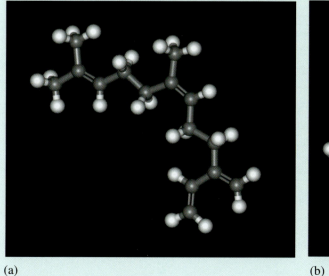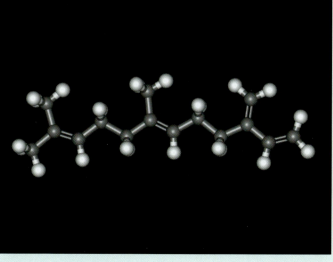

(a)   (b)

Figure 5.5 WebLab ViewerLite images of (a) '*trans*'- and (b) '*cis*'-β-farnesene. 💻

5.1 Some other geometrical isomers

The carbon–carbon double bond is not the only functional group that lends itself to the possibility of geometrical isomers (defined earlier as 'compounds having the same sequence of atoms or groups and differing only in the spatial orientation of these attached groups').

One example is cyclopropane-1,2-diol (see margin).

cyclopropane-1,2-diol

🔵 How can the two alcohol groups be attached differently to the cyclopropane ring? (You may have to use your model kit to answer this question.)

⚪ The three-membered hydrocarbon ring of cyclopropane, is flat, like the carbon–carbon double bond. So the two alcohol groups can be attached on opposite sides, or on the same sides of the plane of the ring (Structures **5.19** and **5.20**, respectively). This is not the case for its positional isomer, cyclopropane-1,1-diol.

5.19 *E*-(*trans*)-cyclopropane-1,2-diol

5.20 *Z*-(*cis*)-cyclopropane-1,2-diol

In just the same way as alkenes, these structures can be designated *E*- and *Z*-, or *cis* and *trans* (if appropriate).

However, even more interesting possibilities are available when we consider atoms or groups around a metal centre. Earlier, we mentioned that many four-coordinate compounds of metals, particularly platinum, palladium and nickel, can have square-planar geometries. This means that they can exist as two possible geometrical isomers, unlike dichloromethane (see Figure 1.4).

Distinct compounds like the two nickel compounds **5.21** and **5.22** can actually be synthesised. Some of these four-coordinate planar compounds have quite different physical properties (colour, melting temperature, magnetic properties, solubility), and can show quite distinct chemical behaviour and reactivity.

5.21

5.22

Perhaps the best-known example of a square-planar platinum complex is the anti-cancer drug, cisplatin (Figure 5.6a and Structure **5.23**). This drug was discovered quite unexpectedly, by researchers investigating the effect of electric currents on bacterial growth. Platinum electrodes were used in their experiments, which were carried out in a medium containing ammonium chloride. An electrolysis product incorporating platinum (**5.23**) was eventually identified as an agent that could inhibit cell division. By contrast, the *trans* form of $PtCl_2(NH_3)_2$ ('trans'-platin)

5.23 cisplatin

(Figure 5.6b and Structure **5.24**) is found to be both toxic and inactive. Cisplatin loses both of its chloro groups in binding together different nucleotide bases in a single DNA strand. This causes the DNA helix to unwind, which terminates the replication process. It is particularly effective against testicular and ovarian tumours. Although cisplatin might seem like a miracle drug, it does have side-effects that can damage the kidney and spinal cord. This led to the development of a number of 'second-generation' platinum-containing drugs, like carboplatin (Structure **5.25**), in which these side-effects are reduced.

5.24 'trans'-platin   **5.25** carboplatin

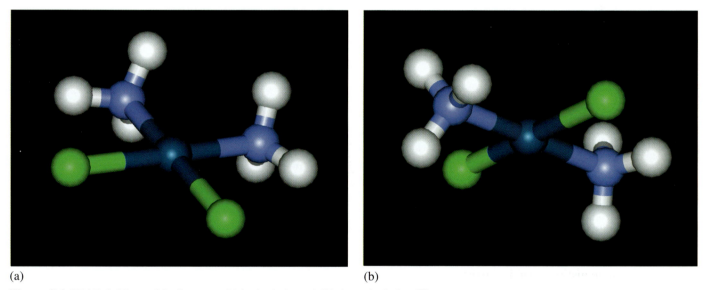

(a)   (b)

Figure 5.6 WebLab ViewerLite images of (a) cisplatin and (b) 'trans'-platin.

With six-coordinate metal compounds, in which the six atoms or groups adopt an octahedral array around the metal centre, even more possibilities exist for geometrical isomerism. Some examples of such compounds are shown in Structures **5.26–5.29**.

5.26   **5.27**   **5.28**   **5.29**

CHIRALITY

6

You may think that by now we have covered all of the possible types of isomerism in organic molecules, but there remains one more type of stereoisomerism, which is the most intriguing and important of all, particularly to living systems. It arises from a fundamental property of the tetrahedral carbon atom — its potential for *chirality*.

Apart from conformational differences, we have only found it possible to make one model of ICH_2CH_2I, and one of CH_3CH_3; similarly, CH_2Cl_2 and even CXY_2Z-type molecules have a unique structure. But what if our saturated carbon atom has four *different* atoms or groups attached?

As before, we shall introduce and develop this concept of chirality through the use of models. In addition to your model kit, you will need a mirror (preferably one that is 12 cm × 12 cm, or larger), and some means of keeping the mirror vertical on your desk or table.

MODEL EXERCISE 6.1
Four different groups attached to carbon

Make a model of CHBrClF exactly as shown in Figure 6.1a. Now mount your mirror vertically, place your model in front of it, and make a model identical with the reflection you can see of your model. This is the mirror image of your first model, in just the same way as a left hand looked at in a mirror appears to be a right hand, and vice versa (Figure 6.1b). Are your two models identical? Can you superimpose them? If not, are there any conformational changes possible for these molecules that would make them identical? What conclusions can you draw from your observations? Do not dismantle the models yet.

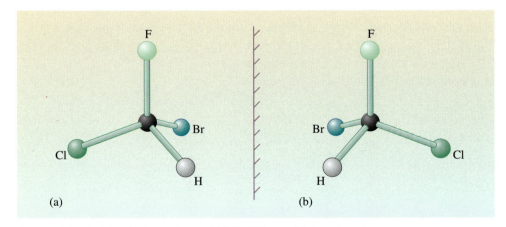

(a)   (b)

Figure 6.1 Models of (a) one form of CHBrClF, and (b) its mirror image.

As with geometrical isomers, these distinct molecules have different configurations. They cannot be interchanged without dismantling the model, in other words without breaking bonds and then remaking them in different positions. Once again, we have a pair of stereoisomers — distinct, separable compounds that have the same molecular formula, with the atoms joined together in the *same sequence*.

Although the bonds to the central carbon atom again have different spatial orientations, unlike geometrical isomers these molecules do not differ in 'shape'; they differ in a characteristic called 'handedness'. The name given to this 'handedness' is **chirality** (from the Greek *kheir* meaning 'hand'). Molecules possessing handedness are called chiral molecules. A carbon atom with four different groups attached to it, wherever it occurs in a molecular structure, is called a **chiral carbon atom**, or a **chiral centre**, although you may also see the term 'stereogenic centre' used instead. Molecules without this property are referred to as **achiral**. A mirror-image pair, like the molecules represented by the models in Figure 6.1, is called a pair of **enantiomers** (from the Greek *enantios* meaning 'opposite'), or an **enantiomeric pair**.

> Enantiomers can therefore be defined as chiral molecules that exist as *non-superimposable mirror images*.

COMPUTER ACTIVITY 6.1 Exploring chirality

In this Activity you will draw chiral molecules in ISIS/Draw, and paste them into WebLab ViewerLite, so that you can manipulate them as three-dimensional representations. Enantiomeric relationships will be explored.

The Activity (in *Exploring the third dimension* on one of the CD-ROMs) should take you approximately 30 minutes to complete.

Enantiomeric compounds cannot easily be distinguished from each other. They have exactly the same melting temperature, boiling temperature, solubility and other physical properties (including identical infrared spectra; see *Separation, Purification and Identification*[1]). They are therefore quite unlike a pair of geometrical isomers, which have distinct physical and chemical properties. Just occasionally, two enantiomers exhibit a feature that can be used to distinguish between them (see Box 6.2, p. 157).

There is, however, one special physical property that allows *all* enantiomers to be distinguished; solutions of individual enantiomers of a pair rotate the plane of **plane-polarized light** in opposite directions. Molecules that rotate the plane of plane-polarized light are called **optically active**, and so enantiomers are sometimes referred to as **optical isomers**. The first person to isolate a pair of optical isomers was Louis Pasteur (Box 6.1).

BOX 6.1 Pasteur and the discovery of chirality

Louis Pasteur (1822–1895, Figure 6.2) was trained as a chemist, but he is probably better known as a biologist, since he could rightly be called the founder of microbiology. He became convinced that living 'germs' were the cause rather than the product of the chemical reactions that occur in fermentation, and by a series of simple but elegant experiments he completely overthrew the then prevalent concept of 'spontaneous generation'. He went further to show that specific micro-organisms were responsible for the production of specific chemical products, and also made the link between 'germs' and infectious diseases. One of his greatest triumphs was the preparation and successful use of a vaccine against rabies. Although partially paralysed by a stroke in 1868, he continued to work, and in 1888 he became the first Director of the internationally renowned Pasteur Institute.

In 1849, he had become interested in the salts of tartaric acid (Structure **6.1**), found in wine, and this led him to crystallize a concentrated solution of sodium ammonium tartrate (Structure **6.2**). Below 28 °C, two different forms of crystal separated out from solution (Figure 6.3). He identified them as being related to each other as non-superimposable mirror images. He laboriously separated the two forms of crystal with a pair of tweezers into 'right-handed' and 'left-handed' piles. He then found that a solution of crystals from each pile showed optical activity when he shone plane-polarized light through them. However, a 50 : 50 mixture of crystals from the two piles was found to be optically inactive. He wrote: 'It cannot be the subject of doubt that [in molecules of the 'right-rotating' tartaric acid] there exists an asymmetric arrangement having a non-superimposable image. It is no less certain that the atoms [molecules] of the 'left-rotating' acid possess precisely the inverse asymmetric arrangement'. This statement was remarkably far-sighted, if you remember that it was not until 25 years later that the proposal of a tetrahedral carbon atom allowed an explanation for chirality at the molecular level to be offered by van't Hoff and Le Bel.

Figure 6.2
Louis Pasteur (1822–1895).

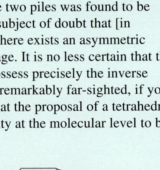

6.1

6.2

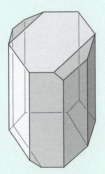 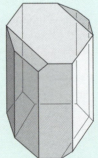

Figure 6.3 Crystals of 'left'- and 'right-handed' sodium ammonium tartrate.

Plane-polarized light can be produced by passing ordinary light of any wavelength through a polarizing material, like the Polaroid sheet used in some types of sunglasses. (A fuller explanation is given in the *Liquid Crystals* Case Study at the end of this Book.) The wavelength usually chosen for experiments is that of the lines at wavelengths around 586 nm in the emission spectrum of sodium atoms (which give sodium lamps their characteristic yellow colour — the so-called sodium 'D lines'). A second sheet of Polaroid, orientated in exactly the same way as the first, will also allow the polarized light to pass through, but if this second sheet of Polaroid is rotated until its plane of polarization is at right-angles to the first, no light emerges at the end of the system. The two pieces of Polaroid, the polarizers, are then said to be crossed. A schematic diagram of the instrument used to measure this rotation — a **polarimeter** — is shown in Figure 6.4.

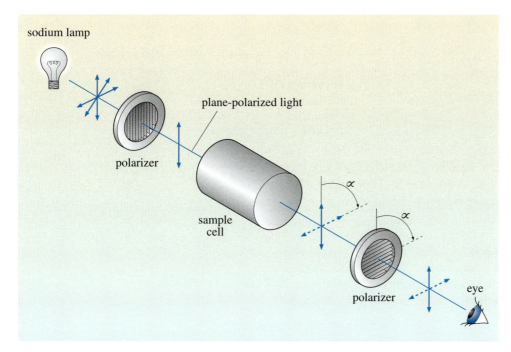

Figure 6.4
Schematic diagram of a polarimeter: from left to right it comprises a sodium lamp, a polarizer (Polaroid sheet), a cylindrical cell containing the sample, and a second polarizer.

If a solution of an optically active compound is placed in the cell between the polarizers, some plane-polarized light now passes through. The optically active molecules in the solution have brought about a rotation in the plane of polarization of the light beam, so the polarizers are no longer fully crossed. If the second polarizer is now turned until all the light from the beam has been extinguished, it is possible to measure the amount of rotation caused by a particular enantiomer.

If, when we look towards the light source, the polarizer has to be turned clockwise to achieve extinction, that enantiomer is labelled (+); the other enantiomer will be (−). A quantity known as the specific rotation, $[\alpha]_D$ (calculated from the concentration of the solution and the path length of the cell containing it) is characteristic for each optically active substance. Some values of $[\alpha]_D$ are given in Table 6.1.

A pure, optically active substance (that is, a sample containing molecules of only one handedness) rotates the plane of plane-polarized light by a characteristic amount. If a particular concentration of one enantiomer rotates the plane of this light

Table 6.1 Sample specific rotations, $[\alpha]_D$, for some chiral molecules; models of all these molecules can be seen as WebLab Viewer files on one of the CD-ROMs (the initial letters R and S are explained later) 🖳

Name	Formula	$[\alpha]_D$
S-(+)-butan-2-ol	$CH_3CH_2CH(OH)CH_3$	+13.9°
S-(+)-lactic acid	$CH_3CH(OH)COOH$	+3.8°
R-(−)-carvone		−60.4°
2R,3R-(+)-tartaric acid	$HOCH(COOH)CH(OH)COOH$	+12.4°
S-(+)-monosodium glutamate	$Na^+ \,^-OOCCH_2CH_2CH(NH_2)COOH$	+25.5°

by $x°$, then exactly the same concentration of the other enantiomer under the same conditions will rotate the plane by exactly $-x°$.

⬤ What does this imply about the rotation caused by a solution containing equal concentrations of two enantiomers?

⬤ There will be *no observed rotation*. The effect of one enantiomer is exactly cancelled by that of the other. A mixture containing equal amounts of two enantiomers is called a **racemic mixture** (from the Latin *racemus* meaning 'juice of the grape').

Unfortunately, the direction of rotation gives us no guide as to the **absolute configuration** — the true orientation in space of the atoms and groups of the molecules causing it. When we use **plus (+)** and **minus (−)** signs, the sign tells us *only* the direction of rotation, and *not* the configuration of the enantiomer.

Just as with stereoisomeric alkenes, the Cahn–Ingold–Prelog nomenclature system (introduced in Section 5) can be used to specify the configurations of chiral carbon atoms; indeed, it was designed principally for this purpose. Using this system, each chiral centre in a molecule can be given an unambiguous label, **R** or **S**.

There is again no connection between R or S and + or −. A compound with an R configuration may rotate the plane of plane-polarized light in either a + or a − direction.

The **priority rules** for specifying absolute configurations are as follows:

(a) Give a priority to each of the four atoms or groups around the chiral atom, by applying the rules given earlier (p. 141) or by consulting Figure 5.2.

(b) Number the atoms or groups 1–4 so that 1 has the highest priority, and the order of decreasing priority is 1 > 2 > 3 > 4.

(c) Arrange the molecule so that the lowest-priority atom or group, 4, is pointing away from you (see Figure 6.5a). It is useful to note that this lowest-priority group is often hydrogen.

(d) With atoms or groups 1, 2 and 3 pointing towards you, let your eye travel from the highest to lowest priority, 1 → 2 → 3 (Figure 6.5b). A clockwise direction of travel is labelled R (from the Latin *rectus*, meaning 'right'), and an anticlockwise direction is labelled S (from the Latin *sinister*, meaning 'left').

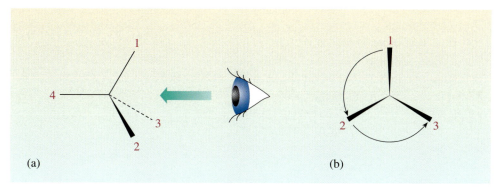

Figure 6.5 (a) The arrangement of groups prior to determining absolute configuration, with the lowest-priority group (4) pointing away from the observer; (b) the molecule as observed in (a). The eye travels anticlockwise from 1 → 2 → 3, so this chiral centre is labelled *S*.

Consider, for example, Structure **6.3**. The rules tell us that Br has priority 1, and H has priority 4. In both the CH_3 and $COCH_3$ groups, the atom bonded to the chiral carbon is another carbon atom, but carbon bonded to oxygen (in $COCH_3$) has a higher priority than carbon bonded to hydrogen (in CH_3). Alternatively, reference to Figure 5.2 shows that a $COCH_3$ group has a higher priority than a CH_3 group.

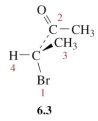

6.3

● Number each group around the chiral carbon atom in **6.3**, and apply the rules to determine the configuration of this molecule.

○ The direction from priority 1 to priority 3 is anticlockwise (Structure **6.4**), so the configuration of **6.3** is *S*.

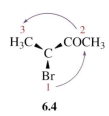

6.4

The absolute configuration of (+)-butan-2-ol (Table 6.1) is *S*-. How can we draw a flying-wedge representation to show its configuration? Firstly, identify the chiral centre. Then draw a 'template' chiral carbon atom with four bonds, and add the hydrogen to the bond pointing into the plane of the paper (as in **6.5**). Now assign priorities to each of the other three groups; for (+)-butan-2-ol, the order of priority is OH (1) > CH_2CH_3 (2) > CH_3 (3). For an *S*-configuration these three groups must be arranged anticlockwise around the carbon centre, in the order 1, 2, 3, as in **6.6**.

6.5

6.6

● The configuration of (+)-lactic acid is also *S*. Draw a flying-wedge representation to show its configuration.

○ Following the guidance given above gives Structure **6.7** for the *S*-enantiomer of lactic acid. The order of priorities is OH (1) > COOH (2) > CH_3 (3) > H (4).

6.7

COMPUTER ACTIVITY 6.2
Absolute configurations of the natural amino acids

Most of the very many chiral molecules that occur naturally are found in only one of their possible configurations. Among the important natural chiral compounds are the amino acids, which are the building blocks of proteins and peptides. They have the general formula $RCH(NH_2)COOH$, so except for the simplest member of the group, glycine, in which $R = H$, they all have four different atoms or groups attached to a central carbon atom.

In this Activity you will draw and examine some amino acid structures in ISIS/Draw, and determine their absolute configurations using the Cahn–Ingold–Prelog priority rules.

This Activity (in *Exploring the third dimension* on one of the CD-ROMs) should take about 20 minutes to complete.

A fascinating observation is that all the naturally occurring amino acids are 'designed' to the same plan; that is, they all have the same configuration. This is true even though in terms of optical activity some are (+) isomers, like alanine and serine, and some are (−) isomers, like phenylalanine and proline.

MODEL EXERCISE 6.2
Fitting amino acids to a chiral template

Build models of glycine, valine, phenylalanine and cysteine in their natural configurations. Each of these amino acid molecules is shown as a model in a Weblab ViewerLite file, which you can use to confirm your structures. In these models, the chiral carbon centre is coloured magenta for easy identification. Now try to fit your four models to the two templates provided in Figure 6.6, by matching hydrogen to 'H', the nitrogen of the amino group to 'N', and the carbon of the carboxylic group to 'C'. The templates given fit models made with the Orbit Kit, but models made with any other model kit should come sufficiently close to a fit to identify the correct template, (a) or (b), unambiguously. Which template do they all fit? Why is glycine the only one that also fits the alternative template?

glycine   valine   phenylalanine   cysteine

What are the configurations (*R* or *S*) of these four amino acids?

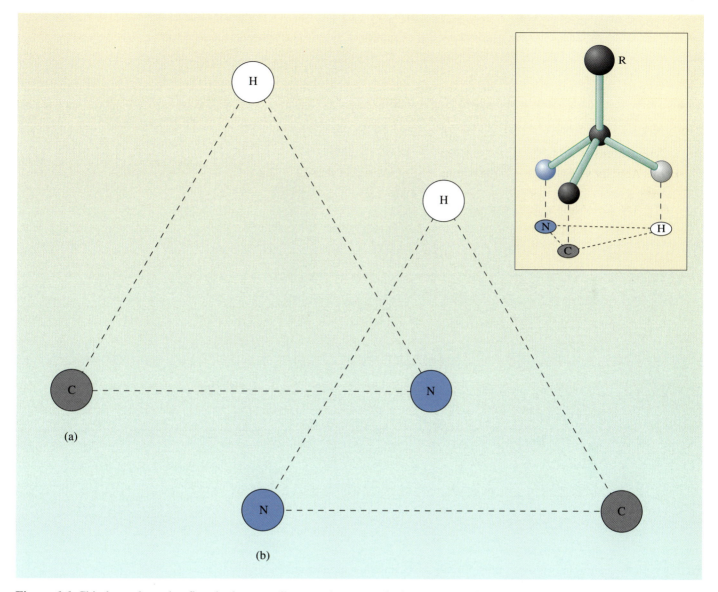

Figure 6.6 Chiral templates that fit only the naturally occurring or synthetic common amino acids. The inset shows the attempted location of a model on a template.

The absolute configuration describes the actual arrangement in space of the groups around a chiral carbon atom. It is not at all easy to determine absolute configurations. The most reliable way is to use X-ray crystallography to establish the exact positions of the atoms. This technique was not readily available until 1951, and by then a great many *relative* configurations had been determined. A series of chemical reactions was used to relate compounds with an unknown configuration to one whose configuration had been arbitrarily chosen. Fortunately, after 1951, when absolute configurations were determined, it turned out that this arbitrary configuration had been the correct one. The question of *representing* configuration is easier; the flying-wedge notation allows configurations to be illustrated simply and exactly.

Not only are most of the physical properties of enantiomers identical, but they frequently behave in the same way chemically as well. In fact, unless there is another chiral agent present, enantiomeric compounds are chemically indistinguishable. An analogy with hands may help to explain why this is (Figure 6.7). Right and left hands do things like picking up sticks, cutting with a knife, throwing a ball, and other activities, in an identical way. When it comes to opening bottles with a corkscrew, putting on gloves, turning on a tap, using scissors, etc., the actions of the two hands are different and distinguishable.

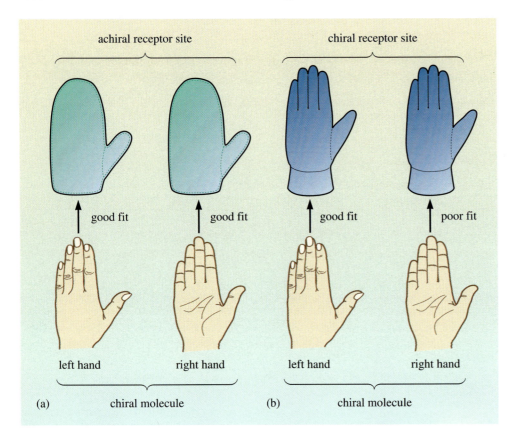

Figure 6.7 Hand-and-glove analogy for the interaction between a chiral molecule and (a) an achiral receptor site, and (b) a chiral receptor site.

So it is with the behaviour of molecules. Enantiomers undergo indistinguishable reactions with non-chiral chemical reagents such as Br_2 and HCl, but with a chiral reagent, significant chemical differences are observed.

We are now able to draw a diagram, similar to Figure E.1 (p. 186), summarizing the features of the two types of stereoisomerism discussed in Sections 5 and 6. This is shown in Figure 6.8.

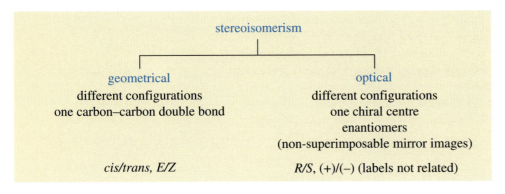

Figure 6.8
Diagram summarizing features of stereoisomerism.

BOX 6.2 Enantiomers in everyday life

Until quite recently, the majority of pharmaceutical products that contain chiral centres were produced as a racemic compound. There are clear disadvantages in doing this. However, since most of the substances that participate in human (and pathogen) metabolism are chiral, optical isomers often exhibit different behaviour. As a consequence, only one isomer is likely to be beneficial; the other may be excreted unchanged, or potentially it could have adverse side-effects. The thalidomide tragedy first drew attention to this problem. Thalidomide was introduced in Europe in the late 1950s, and was prescribed as an anti-nausea drug for treating morning sickness in early pregnancy. It was given as a racemic mixture. Before its disastrous side-effects were recognized, about ten thousand infants world-wide were born with deformed or missing limbs. We now know it is the 'left-handed' form of the thalidomide molecule that is teratogenic (causing deformity in the fetus). However, if only the *R*-enantiomer (Structure **6.8**) had been used, the tragedy would still have occurred, as it is readily converted to the racemic mixture in the body. Yet thalidomide is a remarkable and versatile drug, and almost non-toxic, even to children; provided it is not taken by pregnant women, there is virtually no risk. Indeed, thalidomide is still being manufactured to treat leprosy, and it may have potential for AIDS, arthritis and organ transplant therapy!

6.8 *R*-thalidomide, the 'safe' enantiomer

From the pharmaceutical company's point of view, the knowledge that half your production costs are being wasted is not satisfactory; and the consumer ought to be unhappy about paying for two enantiomers when only one is required. It is possible to separate the two enantiomers, but the procedures are time-consuming, expensive and wasteful. So important has this issue become, that it is true to say that *most* of the synthetic chemistry research currently being carried out, is directed towards the preparation of materials in an optically pure form.

This so-called **asymmetric synthesis** can be achieved in many ways, including the use of chiral starting materials (substrates), chiral reagents, chiral templates, or a chiral catalyst; syntheses using enzymes or other biochemical agents may also be employed. If a chiral molecule is required, a chiral component must be used; otherwise, even in reactions that could give rise to a chiral molecule, a racemic mixture is obtained instead. Since producing single isomers also reduces waste, it is an objective of the 'green chemistry' movement.

We have already seen how insect receptors respond differently to *cis* and *trans* alkenes. The successful interaction of chiral molecules with a whole range of biological receptors also seems to require a quite specific orientation of functional groups, as well as a particular shape, with the implication that receptors too are chiral. In 1872, Lewis Carroll put an interesting question into the mouth of Alice in *Through the Looking Glass*:

> How would you like to live in a looking-glass house, kitty? I wonder if they'd give you milk there? Perhaps looking-glass milk isn't good to drink ...

Milk contains a number of chiral substances, among which is 'milk sugar', (+)-lactose. So, presumably, looking-glass milk would contain (−)-lactose, its mirror image. It is unlikely that looking-glass milk would be harmful, but it would possibly taste differently, and may not be nutritious. (−)-Glucose (the mirror image of natural (+)-glucose) is just as sweet, but it cannot be metabolized by humans. Natural enzymes, accustomed to catalysing the conversion of (+)-glucose into energy, water and carbon dioxide, are unable to handle the 'unnatural' (−)-enantiomer. For this reason it has been proposed as a non-fattening sweetening agent.

By the same token, only the naturally occurring *S*- isomer of monosodium glutamate is an effective flavour-enhancing agent for meats, soups and other foods; its mirror image, the *R*-isomer, does not seem to interact with our taste buds. Even when both isomers occur naturally, they may fit quite different receptors. (−)-Carvone, the *R*-isomer (Structure **6.9**), is the principal odour component of spearmint oil, whereas (+)-carvone, the *S*-isomer (Structure **6.10**), has the odour and flavour of caraway seeds.

6.9 *R*-(−)-carvone   **6.10** *S*-(+)-carvone

QUESTION 6.1

Select from the structures (a)–(d) pairs of molecules that are:

(i)   stereoisomers;

(ii) molecules with the same configuration at the chiral carbon atom;

(iii) different conformations of the same molecule.

You may find it helpful to make models or draw Newman projections.

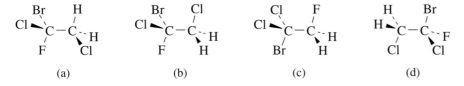

QUESTION 6.2

Which of the following pairs of compounds are enantiomers, and which are different conformations of the same molecule? Again, you will probably find it helpful to make models or draw Newman projections.

6.1 Summary of Sections 5 and 6

1   The alkene fragment is planar and rigid. Rotation about the $C=C$ bond is not usually possible.

2   The necessary condition for alkenes to exhibit stereoisomerism (geometrical isomerism) is that neither of the two alkene carbon atoms should bear two identical groups. If both carbon atoms bear the same two groups, for example H and CH_3, then stereoisomerism is still possible.

3   Geometrical isomers are designated E- or Z- labels according to whether the groups of higher priority in the Cahn–Ingold–Prelog system are on opposite (E-) or the same (Z-) side of the alkene double bond. If one of the groups attached to each carbon of the carbon–carbon double bond is the same, the descriptions *trans* and *cis* may be used.

4   A carbon atom that has four different groups attached to it is called a chiral (or stereogenic) carbon atom (or centre). The configuration of that carbon atom can be specified as R or S; with the group of lowest priority pointing away from the viewer, R is the configuration in which the priorities of the remaining groups decrease in a clockwise direction, whereas S is the configuration in which the priorities decrease in an anticlockwise direction.

5   Tetrahedral molecules CWXYZ exist as two configurations of opposite handedness. The two non-superimposable mirror images are called enantiomers.

6   Equal concentrations of a pair of enantiomers rotate the plane of plane-polarized light equally, but in opposite directions; these are designated (+) and (−). There is no connection between R or S, and + or − labels. A compound with R configuration may rotate the plane of plane-polarized light in either a + or a − direction.

7   The configuration of a molecule can be represented in two dimensions by a flying-wedge projection.

QUESTION 6.3

Which of the molecules (a)–(d) have the same configuration? Try the exercise first without making models; you will find it surprisingly challenging. If you are unable to relate the flying-wedge representations to three-dimensional structures, make models and practise further. Assign one of the labels R or S to each configuration.

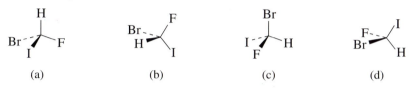

(a)   (b)   (c)   (d)

QUESTION 6.4

Use the sequence rules to specify the configuration of the following molecules:

(a)   (b)   (c)   (d)

MOLECULES WITH MORE THAN ONE CHIRAL ATOM

We shall begin our study of molecules with more than one chiral centre by looking at one of the simplest of such molecules, BrICHCHClF (**7.1**), in which each of the carbon atoms is chiral, but where the two chiral centres are *different*. Chiral carbon centre C-1 bears H, Br, I and CHClF, whereas chiral carbon centre C-2 bears H, Cl, F and CHBrI.

- If we label the configuration of C-1 as 1*R* or 1*S*, and that of C-2 as 2*R* or 2*S*, how many possible combinations are there for Structure **7.1**?

- There are four: 1*R*,2*R*; 1*R*,2*S*; 1*S*,2*R*; 1*S*,2*S*.

- Which of these corresponds to Structure **7.2**? You may find it easier if you build a model, for comparison with the flying-wedge diagram.

- Firstly, rotate the model of **7.2** about the C—C bond to give Structure **7.3**. When the structure is drawn this way (with the hydrogens now pointing away from you), it is easier to see that at carbon centre C-1 the order of priority, I > Br > CHClF, is clockwise, whereas at carbon centre C-2 the order of priority, Cl > F > CHBrI, is anticlockwise. Structure **7.3** is therefore 1*R*,2*S*.

7.1

7.2

7.3

EXERCISE 7.1 Drawing ethanes with two chiral centres

Draw flying-wedge representations for each of the other three stereochemical possibilities for **7.1**. For convenience and comparison, draw each one with the two hydrogen atoms eclipsed, and pointing behind the plane of the paper, as in Structure **7.3**. Remember to include the *R*/*S* designations on your drawings. Try to do this without making models this time; only use models if you get stuck! Remember that an easy way to convert *R* into *S*, or vice versa, is simply to interchange any two of the atoms or groups attached to the chiral carbon atom.

EXERCISE 7.2 Identifying enantiomers

Structures **7.3–7.6** correspond to the four stereochemical possibilities arrived at in Exercise 7.1 (**7.4** corresponds to **E.1**, **7.5** to **E.2**, and **7.6** to **E.3**). From Structures **7.3–7.6**, try to decide which are the *two pairs* of enantiomers. This is not easy to visualize, and you may need to make models to be sure. Now assign *R* and *S* labels to the two chiral carbon atoms, remembering that on reflection, C-2 will be at the right-hand end of the structure, and compare with the *R*/*S* labels given to **7.4–7.6** (**E.1–E.3**) to confirm the identity of the mirror image.

| **7.3** | **7.4 (≡E.1)** | **7.5 (≡E.2)** | **7.6 (≡E.3)** |

If two structures are stereoisomers (as **7.3** and **7.4** (**E.1**) are), but they are *not* superimposable and *not* enantiomers, then we call those structures **diastereomers** (sometimes the term **diastereoisomers** is used). It is important to note that as well as diastereomers being different compounds, unlike enantiomers, they have distinctly different physical and chemical properties. This makes them much easier to separate by conventional procedures, like distillation and recrystallization (these methods are discussed in *Separation, Purification and Identification*[1]).

Now we can describe the stereochemical relationship between the four compounds **7.3**–**7.6**. Pairs of enantiomers are **7.3** and **7.6**, and **7.4** and **7.5**. Pairs of diastereomers are: **7.3** and **7.4**; **7.3** and **7.5**; **7.4** and **7.6**; and **7.5** and **7.6**.

● Have you observed a relationship between *R* and *S* designations, enantiomers and diastereomers?

● The first pair of enantiomers, **7.3** and **7.6**, have the configurations 1*R*,2*S* and 1*S*,2*R*, and the second pair, **7.4** and **7.5**, have the configurations 1*R*,2*R* and 1*S*,2*S*. Each pair of enantiomers has *exactly opposite* configurations at *both* carbon atoms. The difference between diastereomers is that they have only *one* carbon atom of opposite configuration.

● What is the stereochemical relationship between Structures **7.7** and **7.8**? You may need to change the *conformation* of either or both of **7.7** or **7.8** to be sure. Probably making a model will make this easier, but try to do it without first. Manipulating these structures in your head, or on paper, is a skill that you will you need to master.

7.7   7.8

● The two molecules have the same sequence of atoms, and might therefore appear to be stereoisomers. However, viewing Structure **7.7** from the left-hand end, and rotating C-3 by 120° anticlockwise, and viewing C-2 of **7.7** from the right-hand end, and rotating it by 120° clockwise, produces a structure that is *identical* with **7.8**. Alternatively, you may have worked out that both *configurations* are 1*S*,2*R*.

As stated earlier, substituted ethanes undergo internal rotation very readily at room temperature. Their molecules can therefore easily adopt many different conformations. When trying to work out the stereochemical relationship between two structures, therefore, you are allowed (and often may need) to change a given conformation so that you can compare it with another structure. Under normal conditions, different *conformations* of the same molecule cannot be described as different *stereoisomers*, since it is not possible to isolate them separately.

● What is the stereochemical relationship between Structures **7.9** and **7.10**?

7.9   7.10

● The compounds represented by Structures **7.9** and **7.10** are enantiomers. The absolute configurations of both chiral carbon atoms is different in the two structures. If you did not reach this conclusion, make models of the two molecules in the conformations shown (using different-coloured atom centres to represent the different groups). Rotation about the C—C bond shows that the two structures can be shown to be enantiomeric. No amount of rotation will allow *configurations* **7.9** and **7.10** to become identical. Only *conformations* can be interchanged by rotation about single bonds.

● How can you change Structure **7.9** to generate a diastereomer of **7.10**?

● All that is necessary to destroy the mirror-image relationship of **7.9** and **7.10** is to change the stereochemistry at either C-2 or C-3, by interchanging two of the groups attached to one of the carbon centres. Structures **7.11** (change at C-3) and **7.12** (change at C-2) are therefore the two diastereomers of Structure **7.9**.

7.11   **7.12**

In examining Structures **7.3**–**7.6** we have discovered that with *two* different chiral centres in a molecule there are *four* possible stereoisomers, and *two pairs* of enantiomers.

> In general, when a molecule contains *n* different chiral centres, there will be 2^n different stereoisomers possible and therefore half that number, (2^{n-1}), pairs of enantiomers.

We can now return to the first molecule we considered, BrICHCHClF (**7.1**), and use it to confirm this rule. Structure **7.1** has *two* chiral centres, and so will have $2^2 = 4$ possible stereoisomers and $2^{2-1} = 2$ pairs of enantiomers, as we discovered in Exercises 7.1 and 7.2.

COMPUTER ACTIVITY 7.1
The stereochemistry of two sugar molecules

In this Activity you will investigate the stereochemical relationship between the structures of two naturally occurring sugar molecules, by using ISIS/Draw to display the structures in flying-wedge notation, and WebLab ViewerLite to show their three-dimensional nature.

The Activity (in *Exploring the third dimension* on one of the CD-ROMs) should take about 25 minutes.

Many naturally occurring molecules possess a considerable number of chiral centres, each one with a specific configuration. This means that the natural product is only one out of many possible stereoisomers. This is why the structure determination and synthesis of natural products is so fascinating and challenging to organic chemists. The configuration of each chiral centre has first to be determined, and then a synthesis has to be devised to put each chiral centre in place with its correct stereochemistry. When we look at a comparatively simple natural product — for

7.13 oestrone

example, oestrone (a female hormone) — we can see there are four different chiral centres in the molecule (denoted by red asterisks in Structure **7.13**). This means that in theory there are 2^4, or 16 different stereoisomers, comprising eight pairs of enantiomers. However, only *one* of these has hormonal activity.

It is important to be able to identify the chiral centres in a complex molecule, especially when it is presented as a 'framework' structure. Remember that all four valencies of carbon must be satisfied, and so hydrogen atoms may have to be added. In only one of the four chiral centres of oestrone are all the attached groups shown.

QUESTION 7.1

Identify the chiral centres in the perfume and flavouring molecules whose structures are shown as Structures **7.14–7.16**. For each compound, say how many stereoisomers are possible, and how many pairs of enantiomers. Don't forget that unless four bonds to a centre are shown, hydrogen atom(s) have to be added mentally.

7.14 hydroxycitronellal (lily of the valley)   **7.15** ionone (violets)   **7.16** menthol (mint)

Now let us return to simpler compounds, and look at molecules containing two *identical* chiral centres.

MODEL EXERCISE 7.1 *Meso* compounds

Make models of the four stereoisomers of 1,2-dichloro-1,2-difluoroethane (ClFCHCHClF), in the conformation with both hydrogen atoms eclipsed (Structures **7.17–7.20**). Work out the designation, *R* or *S*, for each chiral centre.

7.17   **7.18**   **7.19**   **7.20**

Compare the four models very carefully; what do you find?

So, the 1*R*,2*S* and 1*S*,2*R* configurations of this particular molecule are identical. They are *superimposable mirror images*, which means that the compound represented by this structure is not chiral. We have here an example of a compound whose molecules *contain chiral centres but are not themselves chiral*; such a compound is called a ***meso* compound**. As the molecules of a *meso* compound are not chiral, it will *not* rotate the plane of polarized light; only chiral compounds can do that. *Chirality is a property of a molecule as a whole, and is not simply dependent on the presence of chiral centres.*

Meso compounds can be recognized by the presence of an **internal mirror plane of symmetry**. The mirror plane for 1*S*,2*R*-ClFCHCHClF (Structure **7.21**) runs through the centre of the C—C bond, perpendicular to the plane of the paper, as shown.

The molecule would have *exactly* the same appearance if either half were to be removed and replaced by a mirror in this plane. You can verify this for yourself by

7.21

making half of the **7.21** molecule — by cutting the C—C bond straw in half — and holding the model up to your mirror. It is a sufficient condition for a molecule to be *meso* if it contains a mirror plane in only one conformation; it is usually quite easy to decide which is the best conformation to look for a mirror plane.

The relationship of the *meso* forms **7.18** and **7.19** to the enantiomeric pair **7.17** and **7.20** is also interesting. As they are not mirror images of either **7.17** or **7.20**, the *meso* form must be in a diastereomeric relationship to both. All the other diastereomers we have examined in this Section have shown optical activity; the *meso* form alone is optically *inactive*.

Finally, we need to return briefly to *geometrical isomers*. How can we describe a pair of *Z/E* isomers using the stereochemical terms that we have used for chiral compounds? Perhaps it will be worth using your model kit to make a simple alkene, like but-2-ene, in its *Z* (**7.22**) and *E* (**7.23**) configurations, before going further. Examine the two models.

7.22   7.23

Neither of the compounds they represent can be optically active, since they both possess a plane of symmetry (in the plane of the four carbon atoms), and, if you make the corresponding models, you will find that both are superimposable on their mirror images. (If you cannot superimpose **7.22** or **7.23** on its mirror image, pick up the model, turn it over and try again!) The two compounds **7.22** and **7.23** are *not* mirror images, so they cannot be enantiomers. *These compounds are stereoisomers that are* not *enantiomers: they must therefore be diastereomers*, and like the *meso* forms discussed above, they are optically inactive diastereomers.

7.1 Summary of Section 7

1   Compounds with *n* different chiral centres can have 2^n stereoisomers (2^{n-1} enantiomeric pairs) *unless* a *meso* form (see point 4) is possible.

2   In a molecule that has two different chiral centres, the *R,R* and *S,S* configurations will be a pair of enantiomers, and so will the *R,S* and *S,R* configurations.

3   Stereoisomers that are non-superimposable and are not enantiomers are called diastereomers. This includes *E,Z* (*cis/trans*) geometrical isomers.

4   Even though a molecule contains chiral centres it is not necessarily chiral overall. A substituted ethane, in which the two chiral centres are superimposable on their mirror images is called a *meso* compound. These compounds are achiral, and they are optically inactive because they have an internal mirror plane of symmetry.

QUESTION 7.2

Draw flying-wedge projections of all of the stereoisomers of tartaric acid, HOOCCH(OH)CH(OH)COOH (**6.1**). Label the compounds as diastereomers, enantiomers or *meso*, and draw in the mirror planes, if any. Which compounds would you expect to be optically active?

STEREOCHEMISTRY OF SATURATED RING COMPOUNDS

8

When discussing geometrical isomerism in alkenes, we emphasized the lack of rotation about carbon–carbon double bonds. This structural rigidity is also to be found in cyclic compounds with small ring systems, like the ones in Section 5.1.

8.1 Three- and four-membered rings

We shall use substituted cyclopropanes to illustrate some of the stereochemical features of three-membered rings. The parent compound, cyclopropane, is shown in Structure **8.1**, and mono- and disubstituted chloro derivatives in Structures **8.2** and **8.3**, respectively (note the use of an equilateral triangle in the skeletal notation). Although cyclopropane is a highly strained molecule, it is readily synthesised in the laboratory.

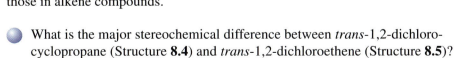

8.1 cyclopropane

8.2 chlorocyclopropane   **8.3** 1,2-dichlorocyclopropane

MODEL EXERCISE 8.1 Monosubstituted cyclopropanes

Make a model of chlorocyclopropane, **8.2**. If using the Orbit Kit, you will need to employ the flexible white straws. It is not possible to construct three-membered rings using the rigid green straws. Other model kits will have their own methods for making small rings. How many isomers of chlorocyclopropane can you make?

MODEL EXERCISE 8.2 Disubstituted cyclopropanes

Make a model of 1,2-dichlorocyclopropane, **8.3**. How many isomers of this compound can you make? What is the relationship between each of them?

The ring structure of cyclopropanes prevents rotation about any carbon–carbon single bond, without first breaking one of the others. There is therefore only one conformation for each molecule. Consequently, stereoisomers are possible in a similar way to those in alkene compounds.

What is the major stereochemical difference between *trans*-1,2-dichloro-cyclopropane (Structure **8.4**) and *trans*-1,2-dichloroethene (Structure **8.5**)?

8.4 *trans*-1,2-dichlorocyclopropane   **8.5** *trans*-1,2-dichloroethene

trans-1,2-Dichloroethene (**8.5**) has a mirror plane running through all six atoms, and so it is superimposable on its mirror image. It cannot therefore have an enantiomer. By contrast, *trans*-1,2-dichlorocyclopropane (**8.4**) has no mirror plane. As you found in Model Exercise 8.2, there are two enantiomeric *trans*-1,2-dichlorocyclopropanes.

QUESTION 8.1

Designate the two chiral carbon atoms in *trans*-1,2-dichlorocyclopropane (Structure **8.4**) as *R* or *S*. If you are not sure which groups are attached to the chiral centre, jot down each adjacent atom, and note all the atoms attached to it. Without starting from scratch, suggest the designation for these centres in the enantiomer, and in the *meso* compound *cis*-1,2-dichlorocyclopropane.

QUESTION 8.2

Use drawings to try to predict the number and configurations of the isomers of 1,3-dichlorocyclobutane. Make models to check your predictions.

8.2 Summary of Section 8

1 Rings restrict rotation about C—C single bonds within the ring, giving rise to stereoisomerism.

2 1,2-Dichlorocyclopropane has three isomers, an enantiomeric pair of *trans* isomers and a *meso cis* compound.

3 1,2-Disubstituted cyclobutanes are similar, but both 1,3-disubstituted cyclobutanes have a plane of symmetry and are achiral.

CONCLUSION

In conclusion, here is a useful algorithm to enable you to identify the stereochemical relationship between isomers (Figure 9.1).

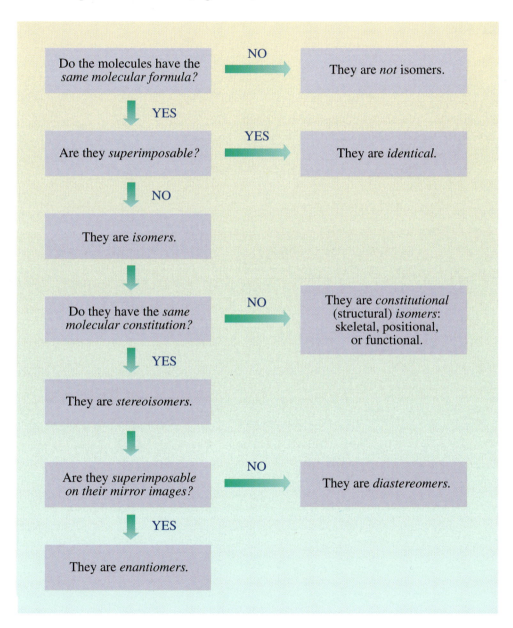

Figure 9.1 Algorithm to determine stereochemical relationships.

QUESTION 9.1

Use the algorithm in Figure 9.1 to identify the stereochemical relationship between the 2-methylpent-3-enoic acids **9.1** and **9.2**. Designate the configurations at the chiral carbon centres and the double bond in each molecule.

Part 2 of this Book has concentrated on describing the stereochemical relationships within and between molecules. Now it is possible to use this knowledge in the understanding of chemical processes. Together with the functional group approach, stereochemistry provides a very sound basis for tackling some of the problems encountered in organic chemistry.

APPENDIX
A BRIEF SURVEY OF ORGANIC FUNCTIONAL GROUPS

All organic molecules, except the saturated hydrocarbons, consist of a carbon framework (or skeleton) with attached functional groups. This way of describing compounds also forms the basis of systematic organic nomenclature. Organic chemistry is principally the chemistry of **functional groups**, which can be defined as the reactive components of an organic molecule. They are important, because almost every organic reaction takes place at a functional group, either to convert it into an alternative functional group, or to extend or reduce the carbon skeleton in some way. Functional groups are characterized by unsaturation and **heteroatoms** (that is, atoms other than carbon and hydrogen); very often, both these characteristics appear in the same functional group. In living systems, as in organic chemistry in general, the most abundant elements are carbon, hydrogen, nitrogen, oxygen, sulfur and phosphorus (the phosphorus is usually found in phosphates, such as $(RO)_3PO$ and pyrophosphates, $(RO)_3POP(OR)_3$). The most common functional groups are listed in Table A.1. This is not a Table for you to learn, but you will find it very useful for reference. As your understanding of organic chemistry grows, these groups will become familiar, and your need to refer to Table A.1 will diminish.

You can see from the Table that some functional groups occur in 'families'. We have, for example, collected together several of the **carbonyl compounds** (those that contain C=O), by listing aldehydes, ketones, carboxylic acids, esters and anhydrides under one heading.

● Which other carbonyl compounds are included in Table A.1?

● Amides and acid chlorides.

As carbonyl compounds all have the C=O group in common, we could hazard a guess that the presence of that one functional fragment would confer a similar reactivity on all members of the series. This sort of generalization is only valid up to a point, but it can help in understanding the chemistry of carbonyl compounds. However, we must bear in mind that some of the characteristic carbonyl behaviour will be affected if there is another functional group nearby in the molecule. For example, the ester function —COOR contains both C=O and OR groups, but it has quite different chemical properties from both ethers (R^1OR^2) and ketones ($R^1R^2C{=}O$). Make sure you do not confuse esters and ethers; they are completely separate classes of compounds. Also, the carboxylic acid functional group contains both C=O and OH groups, but it shows quite distinct behaviour from either ketones or alcohols. The reactions of carboxylic acids can only be understood by considering the COOH group as a whole.

We said earlier that many functional groups are characterized by unsaturation.

● What do we mean when we say that a compound or group is **unsaturated**?

Table A.1   Some common functional groups

Name	Functional group	General formula	Abbreviated formula
Those containing C and H only			
alkene	C=C	(R^{1}R^{2}C=CR^{3}R^4 drawn)	R^{1}R^{2}C=CR^{3}R^4
alkyne	—C≡C—	R^1—≡—R^2	R^{1}C≡CR^2
aromatic or benzenoid compound	(benzene ring)	(benzene ring with R)	C$_6$H$_5$R
Those containing C, H, and O, excluding carbonyl compounds			
alcohol	—OH	R—OH	RCH$_2$OH
diol	HO OH —C—C—	HO OH R^1—C—C—R^4 R^2 R^3	R^{1}R^{2}C(OH)C(OH)R^{3}R^4
ether	O	R^1—O—R^2	R^{1}OR^2
Carbonyl compounds (C=O)			
aldehyde	—C(=O)H	R—C(=O)H	RCHO
ketone	C=O	R^1—C(=O)—R^2	R^{1}COR^2
carboxylic acid	—C(=O)OH	R—C(=O)OH	RCOOH *or* RCO$_2$H
ester	—C(=O)OR	R^1—C(=O)OR^2	R^{1}COOR^2 *or* R^{1}CO$_2$R^2
anhydride	—C(=O)—O—C(=O)	R—C(=O)—O—C(=O)—R	RCOOCOR (rarely R^{1}COOCOR^2)

Name	Functional group	General formula	Abbreviated formula
Those containing C, H and N			
amine	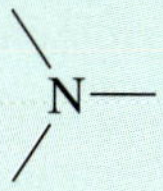 $\diagdown$N—	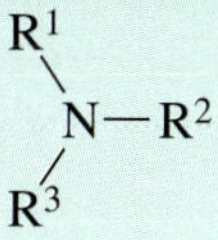 R^1 $\diagup$N—R^2 R^3	$R^1R^2R^3N$
imine	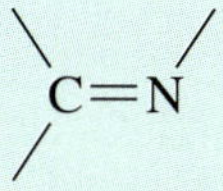 $\diagdown$C=N$\diagup$	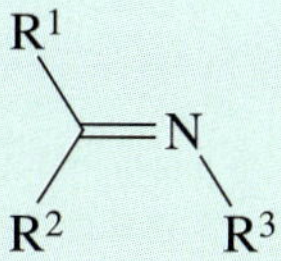 R^1 $\diagup$ C=N R^2 R^3	$R^1R^2C=NR^3$
azo-compound	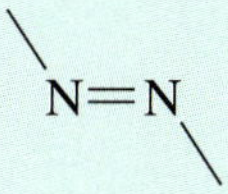 $\diagdown$N=N$\diagup$	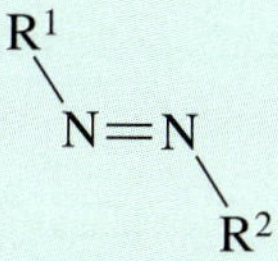 R^1 $\diagdown$N=N R^2	$R^1N=NR^2$
nitrile	—C≡N	R—≡N	RCN
diazonium salt	—$\overset{+}{N}$≡N X^-	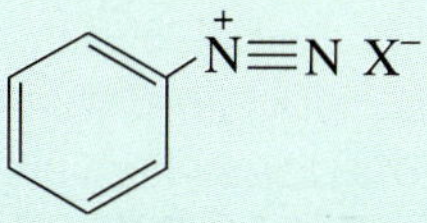 $\overset{+}{N}$≡N X^-	$C_6H_5N_2{}^+ X^-$
Those containing C, H, N and O			
amide	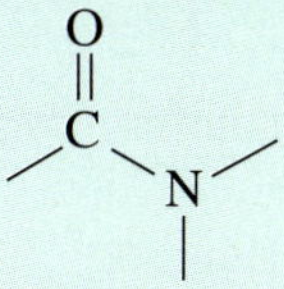 $\overset{O}{\underset{}{\diagup C \diagdown}}N\diagdown$	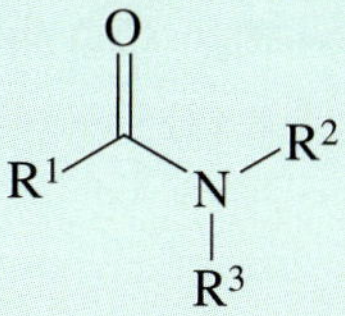 R^1—$\overset{O}{\underset{}{C}}$—$\underset{R^3}{\overset{}{N}}$—$R^2$	$R^1CONR^2R^3$
nitro compound	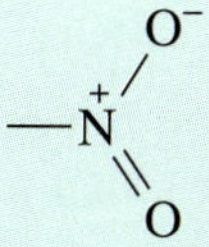 —$\overset{O^-}{\underset{O}{\overset{+}{N}}}$	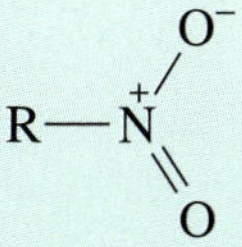 R—$\overset{O^-}{\underset{O}{\overset{+}{N}}}$	RNO_2
nitroso compound	—N=O	R—N=O	RNO
Those containing C, H and other elements			
haloalkane	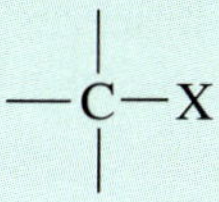 —$\overset{\vert}{\underset{\vert}{C}}$—X	R—X	RX X = F, Cl, Br or I
thiol	—S—H	R—S—H	RSH
acid chloride	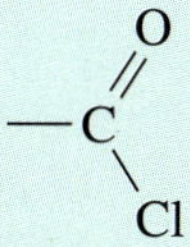 —$\overset{O}{\underset{Cl}{C}}$	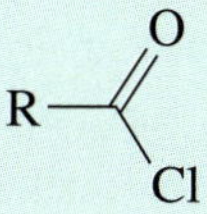 R—$\overset{O}{\underset{Cl}{C}}$	RCOCl
sulfonic acid	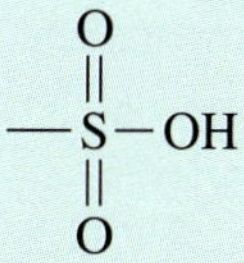 —$\overset{O}{\underset{O}{\overset{\Vert}{\underset{\Vert}{S}}}}$—OH	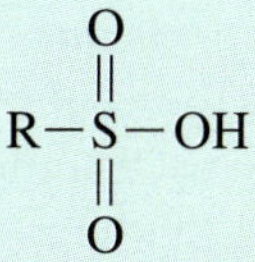 R—$\overset{O}{\underset{O}{\overset{\Vert}{\underset{\Vert}{S}}}}$—OH	RSO_3H (R can be alkyl or aryl)

Looking at Table A.1, you will see that all functional groups except alcohols, amines, ethers, haloalkanes and thiols are unsaturated. The main centres of unsaturation are double bonds between carbon and carbon, carbon and oxygen, or carbon and nitrogen. Sulfur also forms double bonds to oxygen. Carbon–carbon, and carbon–nitrogen triple bonds, as well as nitrogen–nitrogen double and triple bonds are also included in the Table. Although compounds containing benzene rings are formally unsaturated, they are often classified separately as *aromatic compounds*.

● Why do you think that unsaturation causes a functional group to be reactive?

● Unsaturation means that there are multiple bonds present. Therefore more than one pair of electrons is to be found between the two atoms that are bonded together in the functional group. In other words, multiple bonds indicate a concentration of electrons. As reactions depend on the availability of electrons, sites where there are more electrons will be more reactive.

The electrons forming multiple bonds between two atoms are also located further away from the influence of the positively charged atomic nuclei than the electrons of a single bond, and so they are less tightly held. The electrons of unsaturated compounds will therefore be more readily available than those of saturated compounds. This 'ease of availability' is called *polarizability*, and the electrons of multiple bonds are said to be more *polarizable* than electrons in a single bond. You encountered this idea earlier (Part 1, Section 5.4), where it was shown that the large iodide anion is more readily polarized than smaller anions. This occurs because the outermost electrons are further away from the nucleus.

The other characteristic of functional groups is the presence of heteroatoms.

● Why do you think that the presence of nitrogen, oxygen, sulfur, fluorine, chlorine, bromine or iodine causes a functional group to be reactive?

● All these atoms are to be found in the top-right corner of the Periodic Table. Not only do the elements in this region of the Periodic Table have non-bonded pairs of electrons, but they also have a high *electronegativity*, meaning that their atoms attract electrons.

A parameter called the **Pauling electronegativity** (named after the American chemist Linus Pauling; see Box 5.1 of Part 1) can be assigned to each atom to indicate its relative electron 'pulling power'. The absolute value of the numbers is not important here. Simply note that the higher the number, the greater the electron-withdrawing potential. Some Pauling electronegativity values are given in Table A.2. This shows quantitatively what we have said in Part 1, namely that fluorine has the highest electronegativity value.

The importance of electronegativity is seen when atoms are bonded together. Carbon and all the atoms printed in green in Table A.2 form covalent bonds with each other. When an atom of lower electronegativity is covalently bonded to an atom of higher electronegativity, there is an uneven distribution of the electrons in the bond. The more-electronegative atom attracts bonding electrons out of the bond towards itself. This causes the atom of higher electronegativity to become slightly negatively charged, and the atom of lower electronegativity to become slightly positively charged. We indicate this development of charge by the symbols $\delta+$ and $\delta-$. The bond is said to be **polarized**.

Table A.2 Pauling electronegativities of some common elements

			Group			
I	II	III	IV	V	VI	VII
			H 2.1			
Li 1.0	Be 1.5	B 2.0	C 2.5	N 3.0	O 3.5	F 4.0
Na 0.9	Mg 1.2	Al 1.5	Si 1.8	P 2.1	S 2.5	Cl 3.0
						Br 2.8
						I 2.5

● In the covalent bonds between hydrogen and bromine, and between carbon and nitrogen, which atoms bear the $\delta+$ and which the $\delta-$ partial charges?

● The bonds are polarized $\overset{\delta+ \;\; \delta-}{H-Br}$, and $\overset{\delta+ \;\; \delta-}{C-N}$.

● What direction of polarization would you expect to see in **organometallic compounds** (that is, those compounds in which carbon is bonded to a metal)?

● As all the metals in Table A.2 show lower electronegativities than carbon, organometallic compounds will be polarized $\overset{\delta- \;\; \delta+}{C-M}$, where M is a metal.

In the light of what we said earlier, there is one more important observation to make about polarization. The electrons of a double or triple bond are not held as tightly as those in single bonds, and so they will be polarized more readily. This means that the oxygen atom in $C=O$ is more strongly negative, and the carbon atom more strongly positive, than carbon and oxygen in singly bonded $C-O$. Similarly, the nitrogen of a nitrile (with a triple bond between carbon and nitrogen) has a higher $\delta-$ negative charge, and the carbon a higher $\delta+$ positive charge, than the respective atoms in $C-N$.

So this is why heteroatoms make functional groups reactive: through polarization they generate sites of attack for negatively and positively charged reagents (or negatively and positively *polarized* reagents). This also explains why functional groups with both multiple bonds and heteroatoms are generally more reactive than carbon–carbon multiple bonds and singly bonded heteroatomic groups.

Before we leave this brief survey of functional groups, we must mention acids and bases. Hydrogen chloride (HCl) and ethanoic acid (CH_3COOH) are both acids for the same reason, namely polarization! A simple definition of an **acid** is a substance that generates hydrogen ions (H^+) in aqueous solution. Looking at the electronegativities of hydrogen and chlorine, that of chlorine is clearly the greater. Consequently, it will polarize the $H-Cl$ bond $\overset{\delta+ \;\; \delta-}{H-Cl}$. When HCl dissolves in water, this polarization becomes complete, the bond is broken completely, and $H^+(aq)$ ions are formed in solution. Since many reactions are initiated by the H^+ from an acid being transferred to another molecule, it is not surprising that acids came to be defined as *proton donors*.

In ethanoic acid, the hydrogen–oxygen bond is also polarized, $\overset{\delta+ \;\; \delta-}{H-O-}$. So in aqueous solution there is a similar tendency for the O–H bond to break and generate hydrogen ions. However, adjacent to the oxygen of the OH group is the strongly $\delta+$ carbon of the $C=O$ group. The carbonyl group is therefore trying to draw electrons away from the oxygen of the OH group, which makes it all the more active in withdrawing electrons from the O—H bond by way of compensation. This explains why carboxylic acids are much more strongly acidic than alcohols, and also shows how important it is to consider the functional group as a whole, and not just its component parts. However, electronegativity does not provide a complete explanation for acidity, since HCl is a strong acid, and HOH (water!) is a very weak one, even though the electronegativity difference between H and O is greater than that between H and Cl.

Just as an acid may be defined as a proton (H^+) donor, a base may be defined as a proton acceptor. The principal bases in organic chemistry are the amines, RNH_2, R_2NH and R_3N. As nitrogen is more electronegative than either carbon or hydrogen, it withdraws electrons from both C—N and N—H bonds, so that the nitrogen is represented as $N^{\delta-}$; this is a suitable site for attaching H^+. The nitrogen atom is therefore *protonated* to give the ammonium salts, RNH_3^+, $R_2NH_2^+$ and R_3NH^+.

- Why do you think compounds containing the amide functional group ($-CONH_2$) are much weaker bases than the amines?

- In an amide group, the nitrogen atom is adjacent to the $C^{\delta+}$ atom of a carbonyl group. This carbon will therefore strive to withdraw electrons from the C—N bond, making the $\delta-$ charge on nitrogen less. The amide nitrogen is therefore less willing to give its electrons to form a bond with H^+.

It is interesting to note that this electron feedback from nitrogen to carbon also makes the carbonyl group in amides very reluctant to undergo the typical carbonyl reactions of an aldehyde or ketone. Once again, we must be careful not to treat a composite functional group as though it would behave chemically like each of its separate component parts.

Summary

1   Functional groups are characterized by the presence of unsaturation and/or heteroatoms.

2   Composite functional groups do not usually behave chemically in the same way as their component parts acting separately.

3   Bonding electrons are polarizable, especially the less tightly held electrons of multiple bonds.

4   Atoms can be assigned a Pauling electronegativity value, which is a measure of their ability to attract electrons towards themselves by withdrawing them from bonds.

5   Polarization within a bond is indicated by the symbols $\delta+$ and $\delta-$.

6   Acids are proton donors, generating H^+ in aqueous solution; bases are proton acceptors.

LEARNING OUTCOMES

Now that you have completed *The Third Dimension: Molecular shape*, you should be able to do the following things:

1 Recognize valid definitions of, and use in a correct context, the terms, concepts and principles in the following Table. (All Questions)

List of scientific terms, concepts and principles introduced in *Molecular shape*

Term	Page number	Term	Page number	Term	Page number
(+) and (−) labels	152	enantiomeric pair	149	polarized bond	172
absolute configuration	152	flying-wedge notation	128	positional isomer	137
achiral molecule	149	functional group	169	preferred conformation	134
acid	173	functional isomerism	138	priority rules	141
anticlinal conformation	133	geometrical isomer	140	*R* and *S* labels	152
antiperiplanar conformation	133	heteroatom	169	racemic mixture	152
asymmetric synthesis	157	internal mirror plane of symmetry	163	restricted rotation	139
branched chain	136	internal rotation	126	rotation about the carbon–carbon single bond	126
Cahn–Ingold–Prelog system	141	isomer	136	skeletal isomer	136
carbonyl compound	169	*meso* compound	163	skeletal notation	128
chiral carbon atom	149	multiple bond	169	staggered conformation	131
chiral centre	149	Newman projection	131	stereochemistry	128
chirality	149	non-bonded atom	122	stereoisomer	140
cis alkene	140	optically active molecule	149	steric hindrance	135
configuration	139	optical isomer	149	straight-chain hydrocarbon	126
conformation	127	organometallic compound	173	structural isomerism	138
constitutional isomerism	138	oxirane	138	synclinal conformation	133
diastereomer (diastereoisomer)	161	Pauling electronegativity	172	synperiplanar conformation	133
E and *Z* labels	142	pheromone	144	tetrahedral carbon atom	122
eclipsed conformation	131	plane-polarized light	149	*trans* alkene	140
enantiomer	149	polarimeter	151	unsaturated compound	169

2   Use your model kit to solve stereochemical problems effectively. (Questions 3.1, 3.2, 6.1, 6.2, 6.3, 8.2 and all Model Exercises)

3   Distinguish between identical and non-identical molecules, given their Newman or flying-wedge projections. (Questions 3.1 and 9.1, and Model Exercise 3.1)

4   Use ISIS/Draw and WebLab ViewerLite alone, and interactively, for the representation and manipulation of molecular structures. Paste molecular structures generated in ISIS/Draw into WebLab ViewerLite to give three-dimensional representations, and rotate these representations as a whole, or about internal carbon–carbon bonds. (All Computer Activities)

5   Draw Newman and flying-wedge projections from molecular models or accurate descriptions of molecular conformations. (Questions 3.1, 3.2, 3.3, 3.5 and 8.2)

6   Use WebLab ViewerLite to measure bond lengths, bond angles, and non-bonded interatomic distances. (Question 3.4)

7   Recognize, draw, and explain the occurrence of preferred conformations of a substituted ethane. (Question 3.5)

8   Recognize and generate structural (skeletal), positional and functional isomers from a given molecular formula, and differentiate between them. (Questions 4.1, 4.2, 4.3 and 4.4, and Exercise 4.1)

9   Draw and specify the configuration of the stereoisomers of a given alkene, using the descriptors E or Z (and *cis* or *trans* where appropriate). (Questions 5.1, 5.2 and 9.1)

10   Use Newman and flying-wedge projections to distinguish between enantiomers and different conformations of the same molecule. (Questions 6.1, 6.2 and 8.2, and Exercises 7.1 and 7.2)

11   Use flying-wedge representations to show the absolute configuration of a molecule. (Questions 6.1 and 6.3, and Exercises 7.1 and 7.2)

12   Distinguish between flying-wedge projections of enantiomeric configurations. (Questions 6.1, 6.2 and 6.3, and Exercises 7.1 and 7.2)

13   Draw and specify the configuration of chiral carbon atoms as R or S using the Cahn–Ingold–Prelog rules and the priorities in Figure 5.2. (Questions 6.3, 6.4, 8.1 and 9.1, and Exercises 7.1 and 7.2)

14   Draw flying-wedge projections for all the stereoisomers of a compound containing two chiral carbon atoms, and recognize enantiomers, diastereomers and *meso* compounds. (Questions 7.2 and 9.1, and Exercises 7.1 and 7.2)

15   Identify chiral centres in structural and skeletal (framework) representations of molecules. (Question 7.1)

16   Determine the number and types of isomers of a disubstituted cyclopropane or cyclobutane. (Questions 8.1 and 8.2, and Model Exercises 8.1 and 8.2)

QUESTIONS: ANSWERS AND COMMENTS

Question 3.1 (Learning Outcomes 2, 3 and 5)

Conformation (a) is not identical with flying-wedge representation **3.18** because the chlorines are not in the same relative positions.

Conformation (b) actually represents 1,1-dichloroethane, not 1,2-dichloroethane.

Newman projection (c) shows the same conformation of 1,2-dichloroethane as flying-wedge representation **3.18**, if viewed from the left. You could confirm this by building models.

Question 3.2 (Learning Outcomes 2 and 5)

The Newman projections are:

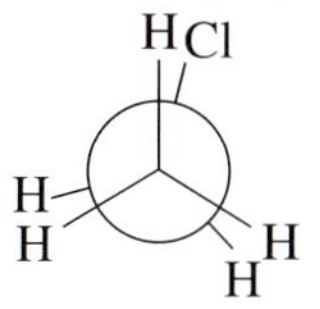

chloroethane 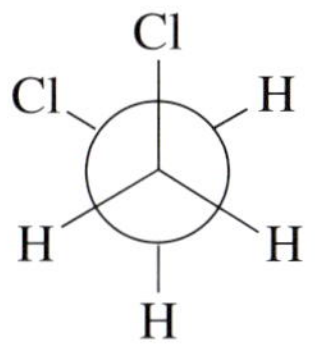   1,2-dichloroethane

The flying-wedge representations are:

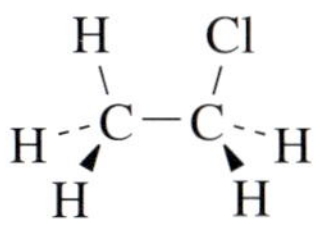

chloroethane 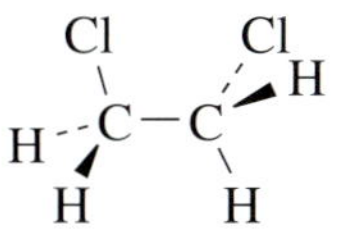   1,2-dichloroethane

Question 3.3 (Learning Outcome 5)

(a)

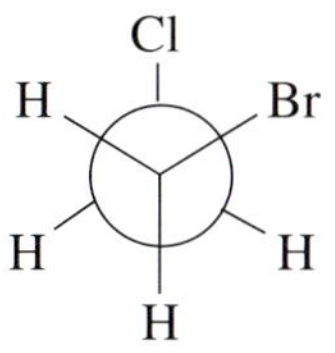

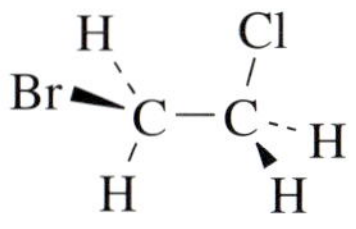

synclinal 1-bromo-2-chloroethane

Other representations with an angle of 60° between the Cl and Br are also correct.

(b)

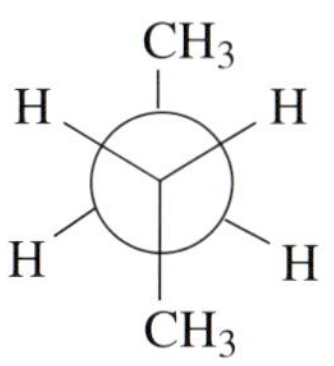

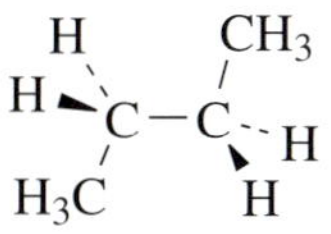

antiperiplanar butane

(c)

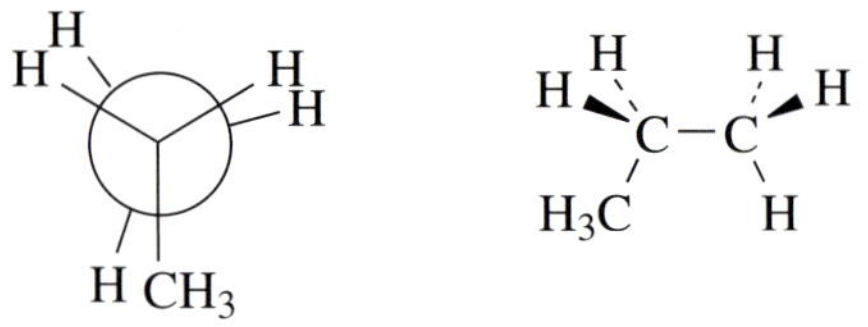

eclipsed propane

(d)

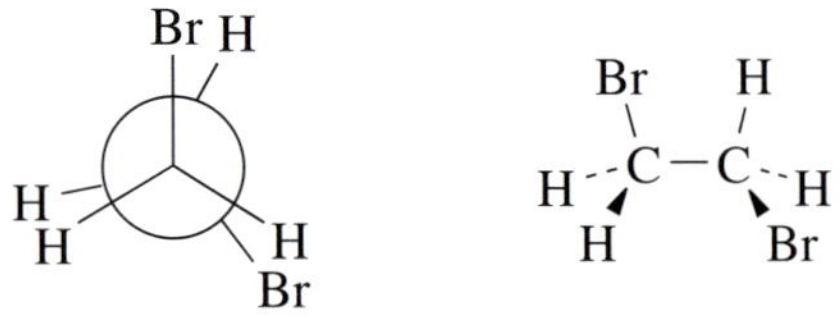

anticlinal 1,2-dibromoethane

(e)

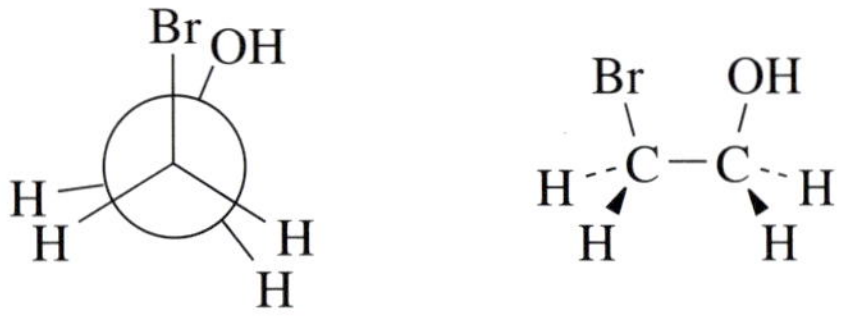

synperiplanar 2-bromoethan-1-ol

Question 3.4 (Learning Outcome 6)

Your measurements may differ slightly from the ones shown.

Conformation	Iodine atoms antiperiplanar	Iodine atoms synperiplanar	Iodine atoms synclinal	Iodine eclipsed with hydrogen
I–I distance/pm	not needed	292	350	449
I–H distance/pm	308	363	310	277
H–H distance/pm	240	220	238	220

sum of van der Waals radii:	I + I/pm	396
	I + H/pm	318
	H + H/pm	240

You should have found the iodine–iodine distance in the synclinal (staggered) conformation to be about 350 pm (≈ 3.50 Å), and in the synperiplanar (iodine–iodine eclipsed) conformation to be about 290 pm. The sum of two iodine van der Waals radii, from Table 1.1, is 396 pm, so even in the synclinal conformation, the two iodine atoms will be squeezed together. This 'squeezing' together of atoms is not the only factor accounting for conformational preference, but it is significant.

Even the sum of the van der Waals radii of iodine and hydrogen (318 pm) is greater than the measured iodine–hydrogen interatomic distances in both the anticlinal (hydrogen–iodine eclipsed, about 280 pm) and the synclinal (about 310 pm) conformations. It is this interaction between the mutually repulsive electron clouds that increases the internal energy of eclipsed conformations (even in the case when hydrogen eclipses hydrogen). The highest level of internal energy for conformations of this molecule, will therefore be found when iodine eclipses iodine (in the synperiplanar conformation).

Question 3.5 (Learning Outcomes 5 and 7)

Applying rules 1 and 2 at the top of p. 135 gives the preferred conformations (a) and (b) below. In (a), the two largest groups, C_6H_5 and Cl are located as far from each other as possible in an antiperiplanar conformation, whereas in (b), the smallest groups (hydrogen) are found to occupy positions between the two larger groups on the adjacent carbon centre, so minimizing the non-bonded interactions. You may have drawn the projections as shown, or some other way, provided they satisfy the same criteria. Remember that, by convention, flying-wedge diagrams are converted to Newman projections by viewing the former from the left.

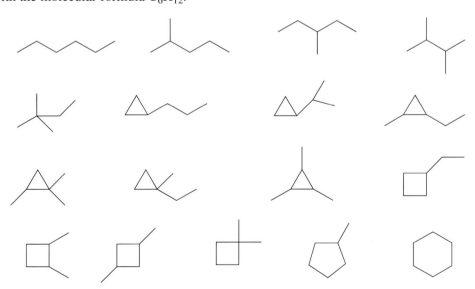

(a)   (b)

Question 4.1 (Learning Outcome 8)

You were probably surprised to see how many structures were possible with even this relatively small number of carbon atoms. The *seventeen* possibilities are shown below. However, not all these skeletons give rise to isomers! If you count up the number of hydrogen atoms required for each of the non-cyclic saturated structures, you will that find *fourteen* are needed in each case, whereas if there is one ring system only *twelve* hydrogens are required. There are therefore only five skeletal isomers with the molecular formula C_6H_{14}, yet there are twelve skeletal isomers with the molecular formula C_6H_{12}.

Question 4.2 (Learning Outcome 8)

There are *twelve* isomers in all with the formula C_4H_8ClBr (shown below). The way to tackle questions like this is first to draw all possible *carbon* skeletons, ignoring the hydrogen, chlorine and bromine atoms; for C_4 there are only two possible skeletons, as we discovered with Structures **4.3** and **4.4**. Then jot down several copies of each skeleton, and arrange the atoms in all possible ways on each skeleton. Finally, check that you have not inadvertently duplicated some structures, and that all structures have the correct number of atoms given in the formula. Note also that every different isomer will have a different systematic name; this is another good way of showing that all your structures are different.

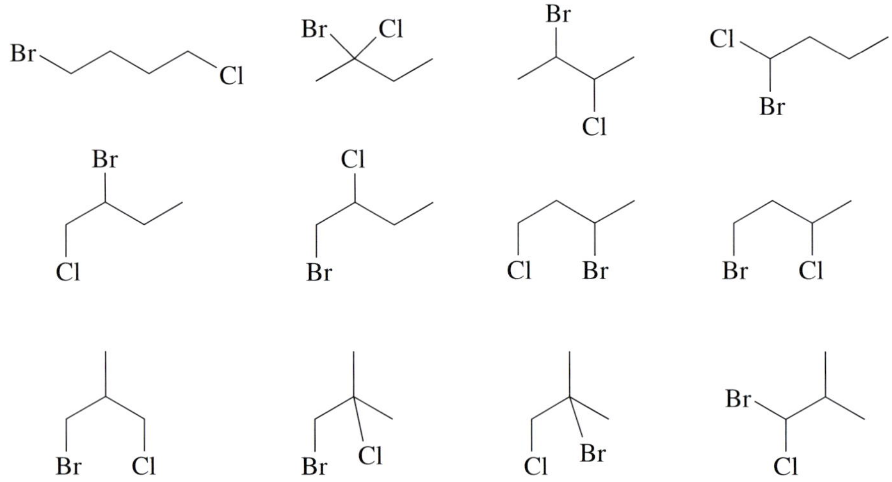

Question 4.3 (Learning Outcome 8)

Four alcohols are possible: butan-1-ol, butan-2-ol, 2-methylpropan-1-ol, and 2-methylpropan-2-ol. Three ethers are possible: 1-methoxypropane, 2-methoxy-propane, and ethoxyethane.

alcohols

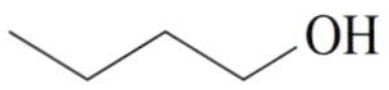

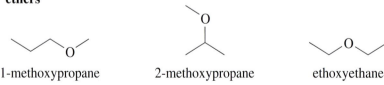

| butan-1-ol | butan-2-ol | 2-methylpropan-1-ol | 2-methylpropan-2-ol |

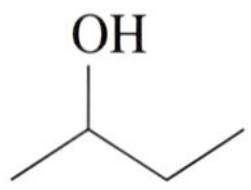

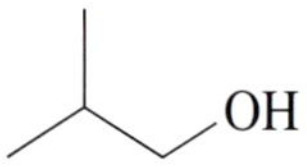

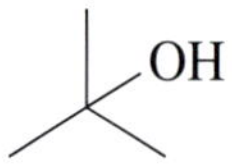

ethers

1-methoxypropane   2-methoxypropane   ethoxyethane

Question 4.4 (Learning Outcome 8)

It is possible to draw many 'paper' isomers, most of which correspond to viable chemical compounds. They are: CH_3CH_2COOH, a carboxylic acid (**Q.1**); CH_3COCH_2OH, a ketone and alcohol (**Q.2**); $HOCH_2CH_2CHO$ (**Q.3**) and $CH_3CH(OH)CHO$ (**Q.4**), both with alcohol and aldehyde groups, so these two are also *positional* isomers; CH_3COOCH_3 (**Q.5**) and $HCOOCH_2CH_3$ (**Q.6**), both esters; CH_3OCH_2CHO (**Q.7**), an ether and aldehyde; the cyclic compounds **Q.8** and **Q.9**, both diols (di-alcohol), and also positional isomers.

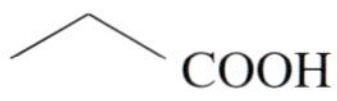

Q.1 propanoic acid 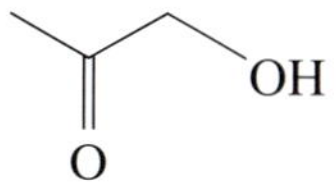   **Q.2** 2-oxopropanol 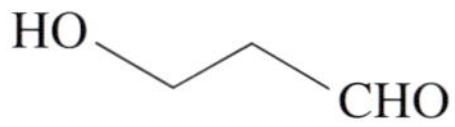   **Q.3** 3-hydroxypropanal 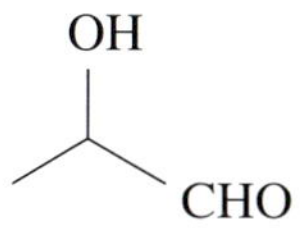   **Q.4** 2-hydroxypropanal

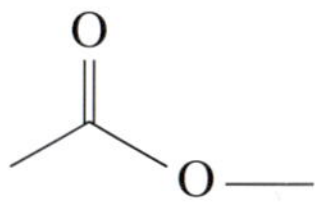

Q.5 methyl ethanoate 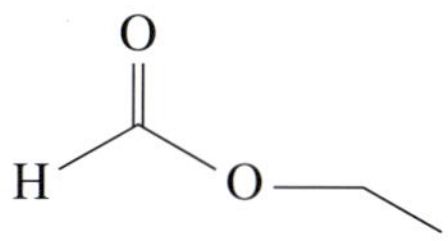   **Q.6** ethyl methanoate 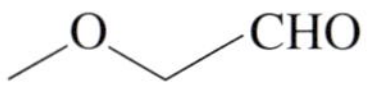   **Q.7** 2-methoxyethanal 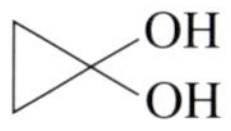   **Q.8** cyclopropane-1,1-diol 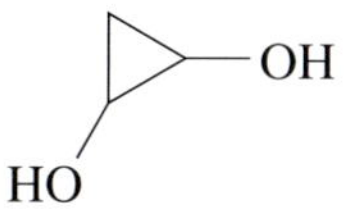   **Q.9** cyclopropane-1,2-diol

There are also some more exotic-looking cyclic hydroxyethers and cyclic diethers:

You may also have included the following structures: $HOCH_2CH=CHOH$ and $CH_2=CHCH(OH)_2$, both of which are diols and alkenes; $CH_2=CHOOCH_3$, an alkene and peroxide. All these are valid structures on paper, but for various reasons most are not likely to be found in a bottle in the laboratory.

Question 5.1 (Learning Outcome 9)

Following the criterion defined on p. 140, 3-phenylpropenoic acid (cinnamic acid) and hex-2-ene exhibit geometrical isomerism, because they fit the general formula, $WXC=CXZ$. 1,2-diphenylethene (stilbene) is a molecule with the general formula $WXC=CWX$, so it also qualifies. The only one of these molecules not meeting our criterion is hex-1-ene. The structures of the pairs of geometrical isomers are shown below, along with that of hex-1-ene.

3-phenylpropenoic acid

1,2-diphenylethene

hex-2-ene

hex-1-ene

Question 5.2 (Learning Outcome 9)

(i) ClHC=CHCl has two isomers:

E (or *trans*)   Z (or *cis*)

(ii) There is only one Cl(CH$_3$)C=CH$_2$, so the labels *cis* and *trans*, or E and Z, are not applicable.

(iii) Similarly, there is only one (CH$_3$)$_2$C=CBrCl.

(iv) (C$_6$H$_5$)HC=CHCH$_3$ has two isomers:

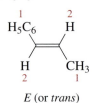
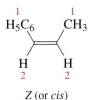

E (or *trans*)   Z (or *cis*)

(v) (C$_6$H$_5$)ClC=CBrCH$_3$ has two isomers, but the labels *cis* and *trans* are not appropriate, whereas E and Z are easily applied.

E (or *trans*)   Z (or *cis*)

Question 6.1 (Learning Outcomes 2, 10, 11 and 12)

(i) Molecules (a) and (d) are stereoisomers, since the atoms are all joined together in exactly the same sequence. They are non-superimposable mirror images, so they are enantiomers. However much we rotate the molecules about the C—C bonds, we can never make (a) and (d) identical. Molecules (b) and (d) are also enantiomers.

(ii) Molecules (a) and (b) have the same configurations around the chiral carbons, since the arrangement of Br, Cl and F atoms is identical. Compound (c) contains the same atoms as (a), (b) and (d), but they are arranged in a different order. Compound (c) is therefore a structural isomer (positional isomer) of (a), (b) and (d) because it has the same molecular formula, C$_2$H$_2$BrCl$_2$F, as the other molecules, but a differ-ent ordering of the atoms. Newman projections of (a), (b) and (d) show clearly that

(a) is the mirror image of (d1), and (b) the mirror image of (d2). Newman projection (d2) is generated by rotating the rear carbon atom of (d1) clockwise through 120°.

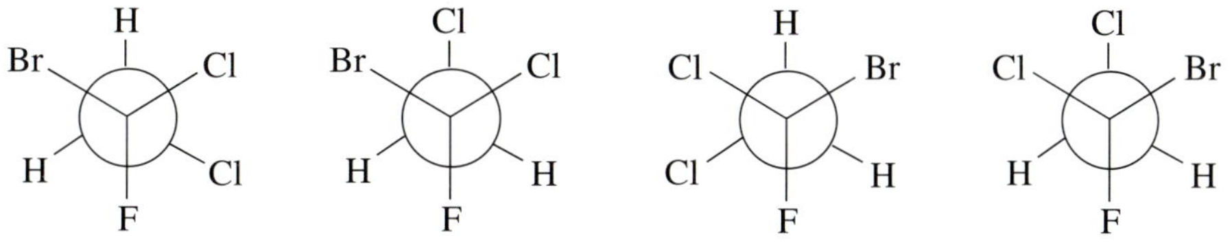

| (a) | (b) | (d1) | (d2) |

(iii) (a) and (b) can easily be made identical by rotation about the C—C bond, and are therefore different conformations of the same molecule.

Question 6.2 (Learning Outcomes 2, 10 and 12)

The pair of molecules labelled (a) can be made identical by rotating the $CHClCH_3$ unit on the right of the left-hand molecule through 120° clockwise, looking from the left. These two are therefore different conformations of the same molecule.

The pair labelled (b) cannot be made identical by any amount of rotation. They are non-superimposable mirror images, and are therefore enantiomers. The left-hand carbon centres are identical, and the right-hand carbon centres are mirror images of each other.

Question 6.3 (Learning Outcomes 2, 11, 12 and 13)

A good way to tackle questions like this is to concentrate on the atoms in the plane of the paper; for example, manoeuvre the 'H' and 'F' of structure (b) into the plane of the paper, in other words into the same positions as the 'H' and 'F' of structure (a). You can then see that for both (a) and (b) the 'Br' is below, and the 'I' above the plane of the paper. Structures (a) and (b) therefore represent the same molecule. Structures (c) and (d) are also different representations of the same molecule, which is the enantiomer of (a) and (b). We need only work out the configuration of one chiral centre, since the others will follow from that. For (a), the priority order is I > Br > F > H, so

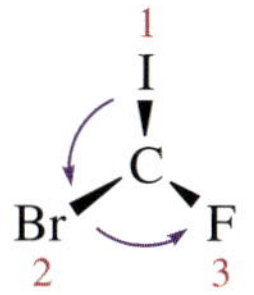

Hence (a) is S-, and so is (b). Structures (c) and (d) therefore have the R- configuration.

Question 6.4 (Learning Outcome 13)

(a)

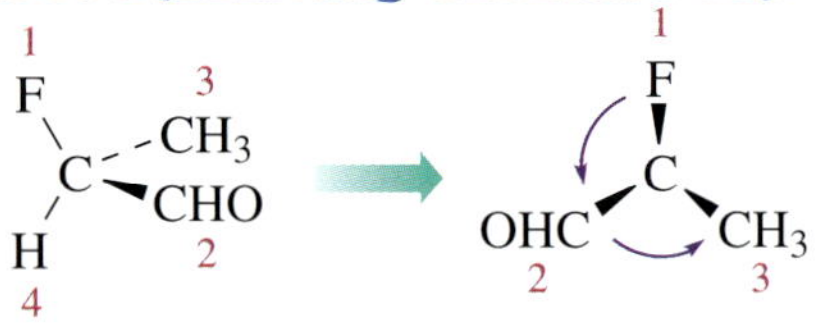

= S

The priority order for this molecule (from Figure 5.2) is:

F > CHO (carbon attached to oxygen and hydrogen) > CH_3 (carbon attached only to hydrogen) > H.

The priorities are shown above for when the molecule is reorientated.

(b)

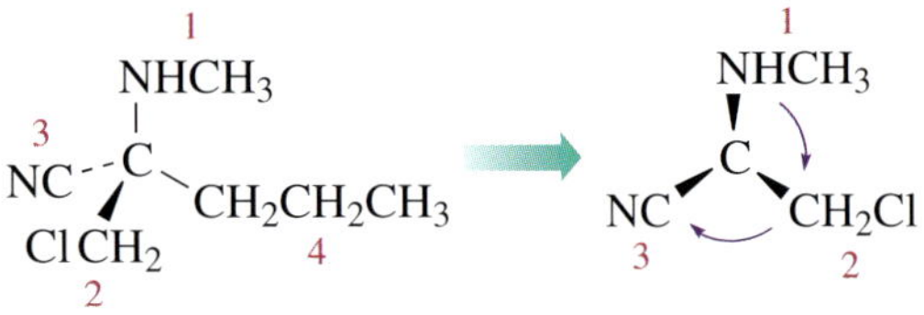

= R

(c)

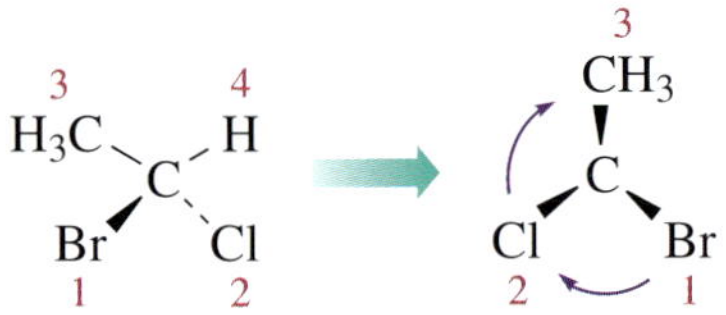

= R

(d)

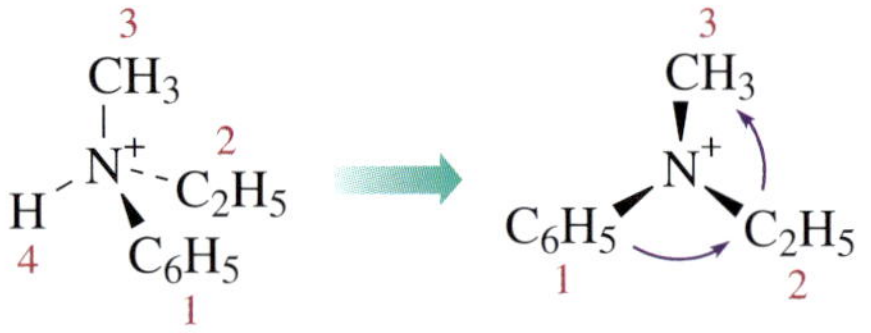

= S

Structure (d) is interesting, in that it shows that not just *carbon* centres with four different groups attached are chiral. Appropriately substituted ammonium ions can also exhibit chirality.

Question 7.1 (Learning Outcome 15)

Hydroxycitronellal (**7.14**) has only one chiral centre; it is carbon-3, and it is attached to methyl, CH_2CHO, a chain of CH_2 groups, and hydrogen (not shown on the structure). Note that although three of the immediately attached groups are CH_2, these are in turn attached to CHO, CH_2 and H, so *all four groups are different*. Two stereoisomers (a pair of enantiomers) could exist.

Ionone (**7.15**) does not have any chiral centres, but the CH=CHCHO side-chain could exist as *Z* and *E* isomers.

Menthol (**7.16**) has three chiral centres, at positions 1, 2 and 5 on the ring. There will therefore be eight (2^3) stereoisomers, and four possible pairs of enantiomers.

Question 7.2 (Learning Outcome 14)

There are two chiral centres in tartaric acid (**6.1**), so there will be 2^2 (that is, 4) possible stereoisomers. Structures (a) and (b) are a pair of enantiomers, and are therefore optically active; both the chiral centres in (a) have a different configuration from the corresponding chiral centres in (b). The third and fourth possibilities are identical, as (c), which is a *meso* compound, and therefore is optically inactive. This is more apparent in the flying-wedge projection of (c) showing the eclipsed conformation (d), where the mirror plane is shown. Structures (a) and (c) are a diastereomeric pair, as are structures (b) and (c).

$$HOOC \quad OH \qquad HO \quad COOH \qquad HO \quad COOH \qquad HOOC \quad COOH$$

$$HO-C-C-H \qquad H-C-C-OH \qquad H-C-C-H \quad \equiv \quad H-C-C-H$$

$$H \quad COOH \qquad HOOC \quad H \qquad HOOC \quad OH \qquad HO \quad OH$$

(a)   (b)   (c) staggered   (d) eclipsed

Question 8.1 (*Learning Outcome 13*)

The order of priority of the three attached groups will be chloro (Cl) > carbon attached to chlorine (CHCl—C) > carbon attached to hydrogen (CH$_2$—C). With hydrogen pointing away from you, the order of priorities at both carbon centres is anticlockwise, therefore both centres in Structure **8.4** will be *S*-. Hence Structure **Q.10** will be described as the 1*S*,2*S*- isomer. The enantiomer **Q.11**, will therefore have the *R,R* configuration, and both *R,S* and *S,R* configurations will be the *meso* compound *cis*-1, 2-dichlorocyclopropane (**Q.12**).

Cl、 H   H、 Cl   Cl、 Cl

H   Cl   Cl   H   H   H

Q.10   **Q.11**   **Q.12**

Question 8.2 (*Learning Outcomes 2, 5, 10 and 16*)

1,3-Dichlorocyclobutane differs in a fundamental way from 1,2-dichlorocyclo-propane. The 1,3-disubstituted cyclobutane has a mirror plane, through the two chlorine atoms, the 1,3-hydrogen atoms, and perpendicular to the ring plane through the broken purple lines in Structures **Q.13** and **Q.14**.

The consequence of this is that the *trans* isomer, as well as the *cis* isomer is superimposable on its mirror image. There are therefore two isomers of 1,3-dichlorocyclobutane, *cis* (**Q.13**) and *trans* (**Q.14**), neither of which is optically active. Neither carbon atom in either **Q.13** or **Q.14** is chiral, since they do not have four *different* groups attached.

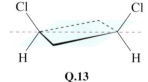

Q.13

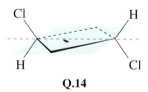

Q.14

Question 9.1 (*Learning Outcomes 3, 9, 13, and 14*)

The molecules have the same molecular formula, but they are not superimposable. They have the same molecular constitution, so they are stereoisomers. Although the chiral centres in Structures **9.1** and **9.2** are mirror images, the arrangement of the substituents on the double bond is different in the two structures. Their mirror images will therefore *not* be superimposable, so they must be *diastereomers*.

The priority of the groups about the chiral carbon centre is —COOH (1), —CH=CHCH$_3$ (2), —CH$_3$ (3), and —H (4). With the hydrogen atom pointing away, as in the structures, the order in **9.1** is anticlockwise, therefore the configuration is *S*; and in **9.2** the order is clockwise, so the configuration is *R*. Both hydrogen atoms are on the same side of the double bond in **9.1**, so the configuration is *Z* (*cis*). They are on opposite sides in **9.2**, so the configuration is *E* (*trans*).

ANSWERS TO EXERCISES

EXERCISE 4.1 (*Learning Outcome 8*)

See Figure E.1.

Constitutional isomerism
same molecular formula, different compounds

skeletal	positional	functional
different structural frameworks	same groups, different position	different functional groups

Figure E.1 Diagram to show the relationship between different types of constitutional isomerism.

EXERCISE 7.1 (*Learning Outcomes 10, 11, 12, 13 and 14*)

There are three possibilities. **E.1** is obtained from **7.3** (the 1*R*,2*S* configuration) by interchanging the F and Cl substituents on C-2. This must therefore be the 1*R*,2*R* combination. Taking **7.3** and interchanging the Br and I substituents on C-1 gives **E.2**, the 1*S*,2*S* combination. The third possibility, **E.3** (1*S*,2*R*), is obtained from **7.3** by interchanging the substituents on both C-1 and C-2.

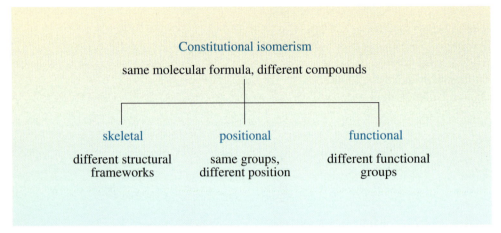

1*R*,2*R*   1*S*,2*S*   1*S*,2*R*

E.1   **E.2**   **E.3**

Each of Structures **7.3** and **E.1**–**E.3** is called 1-bromo-2-chloro-2-fluoro-1-iodo-ethane; they all have the same chemical constitution. But each represents a different compound. Any two compounds that differ only in configuration are stereoisomers, but when we have more than one chiral centre in a molecule, some of the structures will not be enantiomeric with each other.

EXERCISE 7.2 (Learning Outcomes 10, 11, 12, 13 and 14)

7.3 (1R,2S)   **7.3** rotated   **E.3** (1S,2R)

external mirror plane

E.1 (1R,2R)   **E.1** rotated   **E.2** (1S,2S)

Structures **7.3** and **E.3** represent a pair of enantiomers, and so do Structures **E.1** and **E.2**.

When Structures **7.3** and **E.1** are compared, there is no obvious relationship between them; they are not superimposable, and they are not mirror images; they are called *diastereomers*.

ANSWERS TO MODEL EXERCISES

MODEL EXERCISE 1.1 (*Learning Outcome 2*)

No matter how hard you try, you cannot make more than one tetrahedral model of CH_2Cl_2. The molecule is always exactly as shown in Figure ME.1. If you think you have two different arrangements, convince yourself otherwise by rotating one model until it takes on exactly the same appearance as the other. This is true for any molecule with the general formula CX_2Y_2.

With the hypothetical, square-planar carbon atom you should have been able to construct two distinct models, as shown in Figure ME.2. One has the two chlorine atoms and the central carbon atom in a straight line, and in the other the bonds between the two chlorine atoms and the carbon atom form a right-angle. If it were possible to make compounds corresponding to these models, they would be chemically and physically distinct compounds. Note that the tetrahedral model of CH_2Cl_2 can also be displayed using the WebLab ViewerLite software. (However, it is not possible for you to paste in the two square-planar alternatives using ISIS/Draw, since this is 'intelligent' chemical software.)

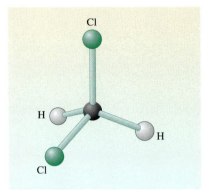

Figure ME.1
Tetrahedral dichloromethane (CH_2Cl_2).

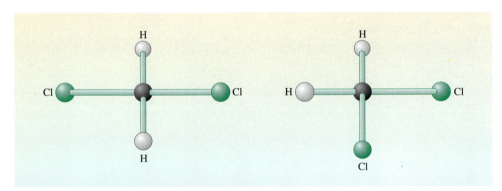

Figure ME.2 Hypothetical square-planar CH_2Cl_2.

MODEL EXERCISE 1.2 (*Learning Outcome 2*)

As for CX_2Y_2, molecules of general formula CXY_2Z (as well as CX_4 and CXY_3) all exist in *only one* tetrahedral configuration. It is impossible to make more than one form of each of these compounds.

MODEL EXERCISE 2.1 (*Learning Outcome 2*)

The odds are very high that the two ethane models you made were not identical.
When we did this exercise, we finished up with the models shown in Figure ME.3.
If you consider our definition for identical molecules, you will see that a mould
made from the first molecular model would not be able to accommodate the second
model.

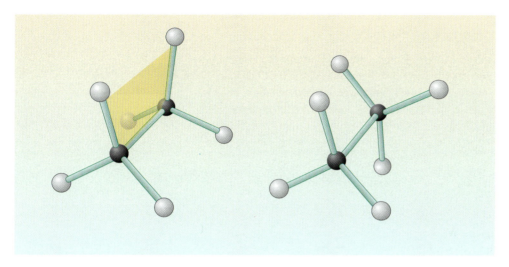

Figure ME.3
Two models of ethane that appear
different owing to different
orientations about the C—C bond.

However, can they be made identical? Conversely, if they are identical, can they be
made non-identical? Examine and manipulate your models so that you can answer
these questions. Do not dismantle the models yet.

MODEL EXERCISE 3.1 (*Learning Outcomes 2 and 3*)

All the following representations (and more) for CH_2BrCl are acceptable, and
correspond to the same molecule.

MODEL EXERCISE 3.2 (*Learning Outcome 2*)

You should not have been able to detect any difference in resistance to rotation in the model as you passed between the various conformations. Nor should you be able to notice any indication during rotation that some conformations are favoured more than others. In these molecular models, no extra effort is needed to achieve one conformation relative to another.

MODEL EXERCISE 3.3 (*Learning Outcome 2*)

The model you should have constructed is a linear zigzag carbon chain, with all the CH_2 groups staggered and antiperiplanar, as illustrated in Figure ME.4. The lowest-energy carbon backbones are the all-staggered, all-antiperiplanar, straight chains. In the polymer poly(ethene), numbers of these chains are able to pack together snugly, which, in part, accounts for the polymer's fibrous nature.

Figure ME.4 Model of a ten-carbon section of an all-staggered, all-antiperiplanar poly(ethene) chain.

MODEL EXERCISE 5.1 (*Learning Outcome 2*)

The first two models that you should have made are shown in Figure ME.5.

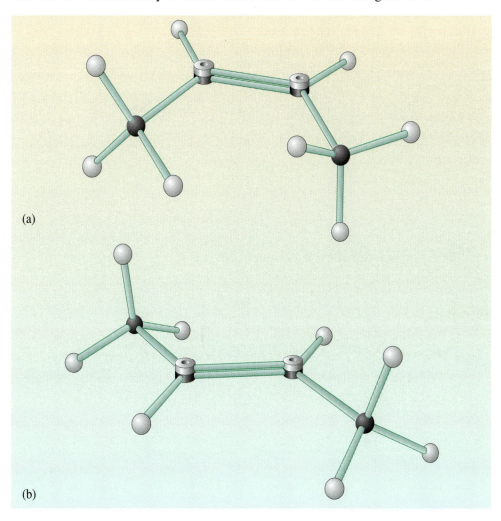

(a)

(b)

Figure ME.5
(a) *Cis*- and (b) *trans*-but-2-ene.

The two models of but-2-ene bear no obvious relationship to each other. They would not fit into the same mould, and so cannot be identical, even though they are constructed from the same components. They can be represented by the formulae **ME.1** (*cis*) and **ME.2** (*trans*).

ME.1   **ME.2**

We would describe **ME.1** by saying that the methyl groups were both on the same 'side' of the double bond, whereas in **ME.2** the methyl groups are on opposite sides.

You should have found that you could only make one model of 2-methylbut-2-ene (discounting the limitation of the non-symmetrical representation of the carbon–carbon double bond).

MODEL EXERCISE 5.2 (*Learning Outcome 2*)

When we tried this exercise, we managed to make a *trans*-cyclooctene, as shown in Figure ME.6, but we decided that the hydrogen atoms crowded the double bond too much. *trans*-Cyclononene seemed more reasonable. In fact, the smallest *trans*-cycloalkene that has been synthesised so far in the laboratory is *trans*-cyclooctene. However, in this molecule the four atoms attached to the double bond are probably not completely in plane with the alkene carbon atoms, owing to the great strain in the ring.

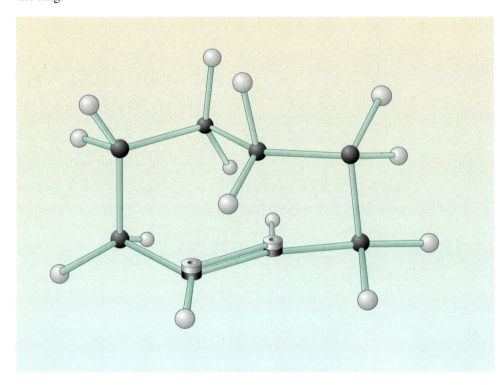

Figure ME.6
A model of *trans*-cyclooctene.

MODEL EXERCISE 6.1 (*Learning Outcome 2*)

If you have constructed the second model of CHBrClF accurately, as in the picture of the mirror-image model in Figure 6.1b, you have as much chance of superimposing these two models as you have of superimposing your right and left hands. It just cannot be done. No matter how much you twist and turn these two molecules they remain distinct, and they cannot be made identical without swapping some of the atom centres, which is equivalent to breaking and then remaking chemical bonds. The only conclusion that can safely be drawn from this experiment is that a compound with four different groups around a tetrahedral carbon atom can exist as two quite different structures that are mirror images of each other.

MODEL EXERCISE 6.2 (*Learning Outcome 2*)

All four of the amino acids fit template (b), in which the sequence N–C–H runs anticlockwise. Glycine fits both templates because it is a non-chiral amino acid. For the other amino acids there are two possible positions for the hydrogen, and only one of them is appropriate to each template. The naturally occurring amino acids nearly all have the *S*-configuration; cysteine is one of the few exceptions.

MODEL EXERCISE 7.1 (*Learning Outcome 2*)

Structure **7.17**, the 1*R*,2*R* configuration, and Structure **7.20**, the 1*S*,2*S* configuration, represent a pair of enantiomers; they are non-superimposable mirror images.

You might have expected **7.18**, the 1*R*,2*S*, and **7.19**, the 1*S*,2*R* structures, to be enantiomers also. However, if you were thorough in your comparison, you would have found that **7.18** and **7.19** were superimposable, and therefore identical structures:

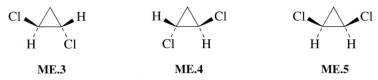

MODEL EXERCISE 8.1 (*Learning Outcomes 2 and 16*)

Only one isomer is possible for chlorocyclopropane. It is superimposable on its mirror image. In fact, there is only one possible monosubstituted compound for all ring sizes, as they all possess a mirror plane. 💻

MODEL EXERCISE 8.2 (*Learning Outcomes 2 and 16*)

ME.3	ME.4	ME.5

There are three possible compounds, **ME.3**, **ME.4** and **ME.5**. For such a simple molecule, the stereochemistry really is quite complex! Compounds **ME.3** and **ME.4** are enantiomers because they are non-superimposable mirror images. If you draw a mirror plane vertically between the two structures, the mirror image relationship can be seen clearly. Each of these compounds is therefore optically active. We usually describe both **ME.3** and **ME.4** as *trans*-1,2-dichlorocyclopropane, by analogy with alkenes, because the hydrogen atoms (and chlorine atoms) are on opposite sides of the ring plane (cf. Structures **5.19** and **5.20** in Section 5.1).

Compound **ME.5** is called *cis*-1,2-dichlorocyclopropane, and it is superimposable on its mirror image. It is therefore a *meso* compound and has a diastereomeric relationship with compounds **ME.3** and **ME.4**. An internal mirror plane of symmetry can be drawn through the unsubstituted ring carbon atom and the mid-point of the opposite carbon–carbon bond.

FURTHER READING

1   L. E. Smart, *Separation, Purification and Identification*, The Open University and the Royal Society of Chemistry (2002).

ACKNOWLEDGEMENTS

Figures

Figure 1.2: Science Photo Library; *Figure 1.5b*: courtesy of Museum Boerhaave Leiden; *Figure 5.3*: courtesy of IACR-Rothamsted; *Figure 5.4*: courtesy of Rothamsted Experimental Station, part of the Institute of Arable Crops Research; *Figure 6.2*: Science Photo Library.

Case Study
Liquid Crystals –
the fourth state of matter

Corrie Imrie

University of Aberdeen

THE DISCOVERY

Towards the end of the nineteenth century, an Austrian botanist, Friedrich Reinitzer, was interested in the role of cholesterol in plants. As part of this research he had synthesised many derivatives of cholesterol, and in 1888 he sat down to measure the melting temperature of cholesteryl benzoate (Figure 1.1). To his great surprise, the sample appeared to have two melting temperatures! At 145.5 °C, the crystals coalesced to form a cloudy fluid, but this suddenly cleared at 178.5 °C. This was a quite remarkable observation, made even more so by what happened on cooling the clear liquid. He noticed that a deep violet–blue colour swept across the sample, leaving behind a turbid liquid, which on further cooling passed through several colour changes before solidifying. Reinitzer had no idea whether these colour changes were the result of physical or chemical changes, and he couldn't explain how one compound could have two melting temperatures. For centuries it had been known that a pure compound should show a single sharp melting temperature.

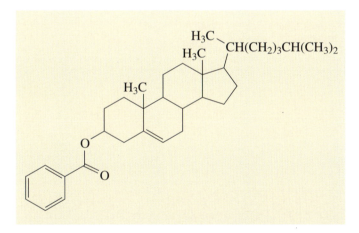

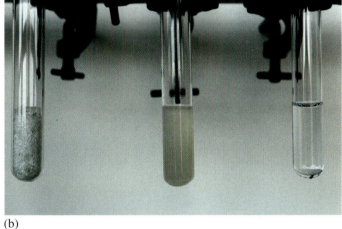

(b)

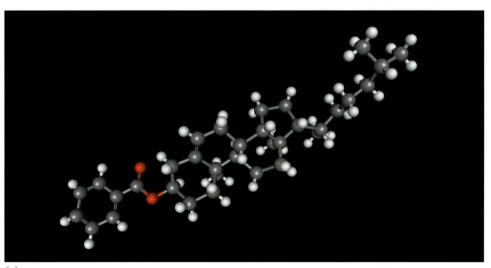

(a)

Figure 1.1
(a) The rod-like molecular structure of cholesteryl benzoate as a line drawing and a WebLab ViewerLite image. (b) Crystals of cholesteryl benzoate, showing a cloudy mesophase; temperature increases from left to right. 🖥

Reinitzer turned to a German physical chemist, Otto Lehmann, for help in explaining this bizarre behaviour. Lehmann's own research involved studying the processes of melting and recrystallization, and one of his main achievements was in developing the polarized light microscope (see Box 1.1). Lehmann used his microscope to study Reinitzer's compound. To his astonishment, he found that the turbid liquid existing between the crystal and clear liquid phases was brightly coloured when viewed between crossed polarizers. Lehmann knew this meant that the turbid liquid was birefringent, a property normally only associated with crystals; but at the same time it also possessed the ability to flow, a liquid-like property. This combination of crystal and liquid-like behaviour led Lehmann to propose that 'crystals exist with a softness being so considerable that one could call them nearly liquid'. He christened such crystals *fliessende Kristalle* or **liquid crystals**, a name that has now stuck for over a hundred years, although our interpretation of the molecular significance of their behaviour has changed somewhat!

The discovery of liquid crystals was only possible through the collaborative efforts of two scientists with very different backgrounds, and this set the pattern for what was to come. As we shall see, the development of liquid crystal science and technology has necessarily involved multidisciplinary research teams involving chemists, physicists, engineers and biologists, and all this long before this way of working became fashionable!

BOX 1.1 The polarized microscope and birefringence

Light consists of two oscillating fields: an electric field and a magnetic field, which oscillate at right-angles to each other and to the direction of propagation of the light (Figure 1.2a). The plane containing the oscillating electric field is called the *plane of polarization* of the light. In general, light will contain many such contributions, with their planes orientated randomly (Figure 1.2b), and it is therefore referred to as *unpolarized* — sunlight is a good example of unpolarized light. It is possible to use a special polarizing filter or *polarizer*, which just allows one plane of polarization through (Figure 1.3a and b), and this light is then called *plane polarized*. (See also the description of the polarimeter in Part 2 of this Book, Figure 6.4.)

Figure 1.2
(a) Light represented as oscillating electric and magnetic fields at 90° to each other.
(b) Unpolarized light has an infinite number of different orientations of the electric field.

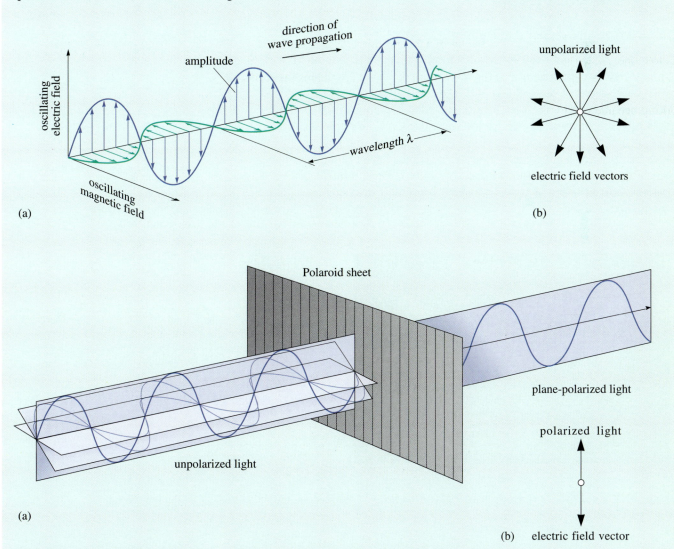

Figure 1.3 (a) Plane-polarized light is produced by passing a beam of unpolarized light through a polarizing filter like a Polaroid sheet. (b) in polarized light, the oscillating electric field lies in a single plane.

As the oscillating electric field of the light passes through a material, it causes the electrons in the atoms and molecules to vibrate. The electric field produced by the vibrating electrons modifies the electric field of the light, and the net effect is to reduce the effective speed of the light in the material; the stronger the electric field in the crystal, the slower the light, which gives rise to the well-known phenomenon of refraction. The extent of refraction is determined by the *refractive index* of the material, which is the ratio of the speed of light in a vacuum to the speed of the light in the material. The molecules in a crystal are arranged in a three-dimensional lattice. It is often the case that some crystal planes contain more molecules than others; such crystals are said to be **anisotropic**. (Cubic crystals are the same in each direction; they are said to be **isotropic**.) Light with the plane of polarization orientated one way, passing through an anisotropic crystal, experiences a different electric field from light with the plane of polarization orientated differently. These two rays therefore travel through the crystal at different speeds.

Plane-polarized light entering an anisotropic crystal can be thought of as consisting of two contributions (or rays), both of which are plane-polarized, but with their planes of polarization at 90° to one another. These two rays will encounter different atomic arrangements with different refractive indices, and thus travel through the crystal at different speeds. If the refractive indices are very different, the crystal is said to have a high **birefringence**, and the two rays may be refracted so differently that it is possible to see two distinct images — so-called **double refraction** (Figure 1.4a and b). Thus, birefringence is a macroscopic property, which is a manifestation of long-range molecular orientational order.

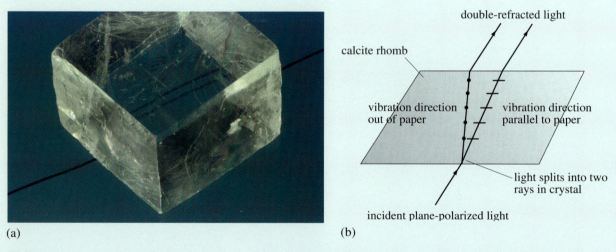

(a)　　　　　　　　　　　　　　　　　　　　(b)

Figure 1.4　(a) Double refraction in a crystal of calcite. (b) In calcite the refractive indices for the two rays are very different.

The polarized light microscope (Figure 1.5) is a conventional microscope, except that the sample is viewed between crossed polarizers. The term 'crossed polarizers' refers to two polarizers arranged such that the plane of polarization of the second polarizer is at right-angles to the first. Plane-polarized light selected by the first polarizer cannot therefore pass through the second polarizer, called the *analyser*. For a sample to appear bright, therefore, it must possess the ability to rotate the plane of plane-polarized light, because only then can light pass through the analyser. The colours observed in samples are due to interference effects.

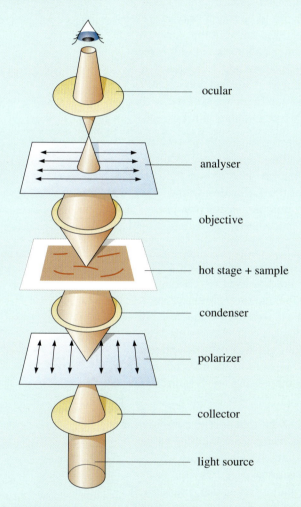

ocular

analyser

objective

hot stage + sample

condenser

polarizer

collector

light source

Figure 1.5 A schematic diagram of a polarized light microscope, widely used in the study of crystals, and the most convenient tool for identifying liquid crystal phases.

THE FOURTH STATE OF MATTER

2

In order to understand Reinitzer's observations at a molecular level, we must first consider the molecular organization found in crystals and conventional liquids. The molecules in a crystal are arranged in a three-dimensional lattice, and are said to have both long-range *positional* and long-range *orientational* ordering. The long-range positional ordering arises because the molecules are to be found only in well-defined lattice positions, on which the molecules are all orientated in the same direction; this leads to long-range orientational ordering. The intermolecular forces that hold the molecules in their three-dimensional lattice are stronger in some directions than others, and so are *anisotropic*. For example, if we consider a pair of rod-like molecules, then the intermolecular forces between them are stronger if they are parallel with respect to each other than if they are perpendicular, in a T-shape.

As a normal crystal is heated, the thermal energy causes the molecules to vibrate more and more, until at the melting temperature these vibrations are sufficiently energetic to overcome the intermolecular forces holding the molecules in place. This causes an abrupt collapse in the long-range ordering of the molecules; in the resulting liquid the molecules are free to move about randomly. A conventional liquid is isotropic and has only short-range positional, and short-range orientational ordering of the molecules.

However, it is easy to imagine two other possible melting processes. In the first of these, thermal energy is sufficient to overcome forces holding the molecules in the same orientations, even though the molecules remain located on their lattice sites. Thus, the molecules are in well-defined positions, but are free to rotate. These are referred to as **plastic crystals**. Alternatively, the melting process may involve the molecules losing their positional ordering, even though the long-range orientational order persists; such phases are termed **liquid crystals**. Further heating eventually destroys the orientational order, giving rise to the second melting temperature, which is referred to as the **clearing temperature**.

We can now begin to understand why liquid crystals exhibit this unique combination of crystal-like properties, such as birefringence, as this stems from the long-range orientational ordering of the molecules. The fluidity, on the other hand, arises because the molecules have moved from the lattice positions. As a consequence, liquid crystals do not fit into the classical division of matter into solids, liquids and gases; instead, they constitute a fourth state of matter. A liquid crystal phase is also referred to as a **mesophase** (from the Greek *mesos*, 'middle), and materials that show liquid crystallinity are called **mesogens**.

TYPES OF LIQUID CRYSTALS

3

We have seen that liquid crystals are orientationally ordered fluids. Within this generic description exists a wide range of liquid crystal phases or mesophases, which may be distinguished by the degree of positional order they possess. Before we describe the molecular organization found within these various mesophases, we must first consider the types of molecules that exhibit liquid crystallinity. For the time being, we shall restrict ourselves to materials that are liquid crystals below a particular temperature; these are known as **thermotropic liquid crystals**.

The overwhelming majority of thermotropic liquid crystals are composed of molecules with a rod-like shape; these are referred to as **calamitic liquid crystals** (from the Greek *calamus*, 'reed' or 'cane'). It is predicted that for a molecule to be liquid crystalline its length should be at least three times greater than its breadth. Only with such a shape will long-range orientational order persist above the melting temperature of the crystal. It is easy to understand why this is the case by filling a box with pencils and shaking it. After shaking, the pencils will all lie in the same direction, but their relative positions will be random, just like a liquid crystal phase! This rod-like molecular shape is most often achieved by attaching one or two alkyl chains to a rigid core consisting of benzene rings separated by short unsaturated linkages (see Figure 3.1). The intermolecular forces between these rigid units are strongest when the molecules line up parallel to each other. This rigid section of the molecule is referred to as the 'mesogenic unit'. The role of the alkyl chains is largely to reduce the melting temperature of the compound, but they are also used to change the type of liquid crystal phase. We shall return to a more detailed discussion of how molecular structure affects liquid crystallinity later, but in the meantime we need to consider how these rod-like molecules are organized in liquid crystal phases.

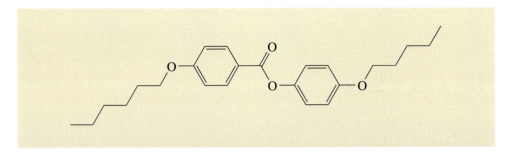

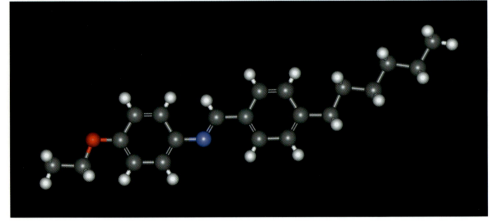

(i)

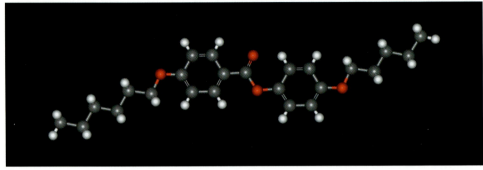

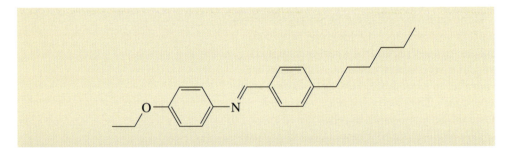

(ii)

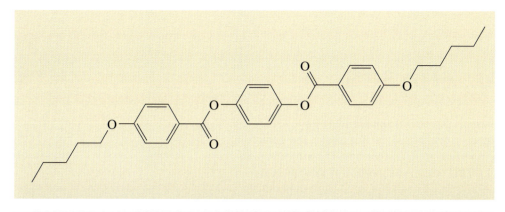

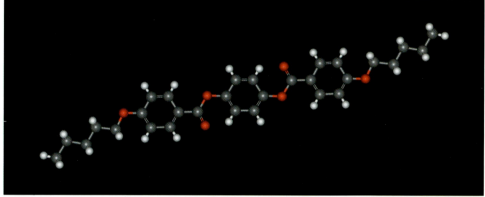

(iii)

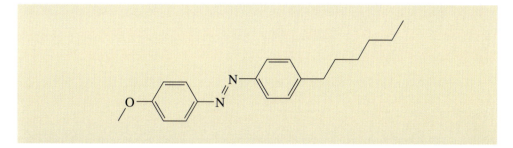

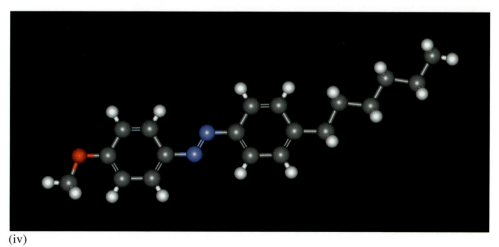

(iv)

Figure 3.1 Line drawings and WebLab ViewerLite images of some typical rod-like or calamitic liquid crystals.

LIQUID CRYSTAL PHASES

4

The **nematic phase** (from the Greek *nema*, 'thread') is the simplest and technologically the most important liquid crystal phase. In this phase, the centres of mass of the molecules are distributed randomly, but, on average, the molecules all point in the same direction (Figure 4.1). This common direction is known as the **director**.

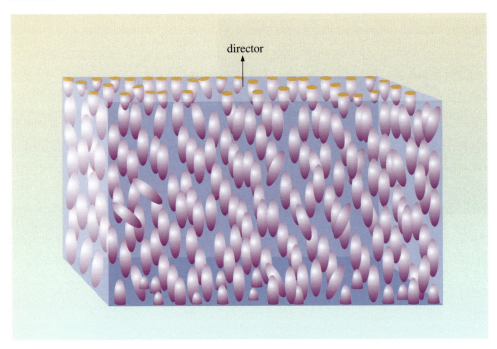

Figure 4.1 A sketch of part of the nematic phase of a liquid crystal.

There are many other types of liquid crystal phases in which the common structural feature is that the molecules are arranged into layers. These are called **smectic phases**, and are differentiated by the arrangement of the molecules within the layers. The various smectic phases are referred to using a single letter (these were assigned chronologically in the order in which the phases were discovered).

The simplest smectic phase is the smectic A phase, in which the molecules are arranged randomly within the layers, there are no correlations between molecules in adjacent layers (Figure 4.2a), and the principal axis of the molecules is, on average, perpendicular to the layers. Smectic phases in which the molecules tilt are also common; the simplest of these is the smectic C phase, which is the tilted analogue of the smectic A phase (Figure 4.2b). It is important to note that the extent of the layering in these phases is always greatly exaggerated in representations such as those in Figure 4.2. In reality, the layered structure is often very weak, and can only be detected using techniques such as X-ray diffraction.

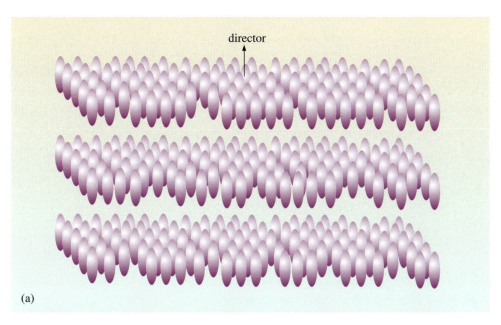

(a)

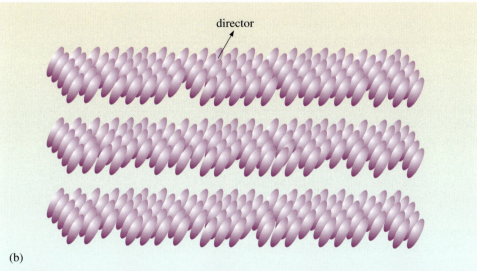

(b)

Figure 4.2 Sketches of the molecular organization within (a) the smectic A phase, and (b) the smectic C phase.

Many smectic phases possess a higher degree of positional order, where, for instance, the molecules *within* a layer may be ordered (smectic phases B, F and I).

IDENTIFYING LIQUID CRYSTAL PHASES

5

We have seen that there is a large number of different liquid crystal phases, and so the question is, how do we know which is which? More than a century after playing a central role in the discovery of liquid crystals, the polarizing microscope remains the most widely used method of characterizing liquid crystal phases. Each mesophase shows characteristic patterns when viewed through the polarizing microscope. These patterns or optical textures can be used like a fingerprint for the phase in question.

The optical textures of mesophases typically consist of a coloured background arising from the birefringent nature of liquid crystals, superimposed on which are dark lines, 'brushes' or points, which result from defects or *disclinations* in the phase. These disclinations are analogous to dislocations in a crystal, and occur where the orientation of the director changes abruptly. Thus, in a bulk sample the director distribution is not normally uniform. Instead there are domains, each with its own director, arranged at various angles with respect to each other. Scattering of light at the interfaces between these domains give rise to the cloudy appearance of the liquid crystals that first caught Reinitzer's eye (Figure 1.1b). The coloured background indicates that the directors lie at some angle with respect to the axes of the polarizers. If the directors in a region lie either perpendicular or parallel to the polarizers' axes, then the field of view appears black; they are called *homeotropic* or *pseudo-isotropic*.

Some characteristic optical textures are shown in Figure 5.1. The nematic phase (Figure 5.1a) shows a texture in which dark areas converge at point disclinations, where the director is essentially undefined. This is a very characteristic texture and is called a *Schlieren texture* (from the German word *Schliere* meaning 'streak'). The texture observed for a smectic A phase (Figure 5.1b) contains fan-like coloured regions and homeotropic areas (black). When increasing temperature causes a phase transition to the smectic C phase, the optical texture shown in Figure 5.1c is obtained. The appearance of the coloured regions changes, and the black area is replaced by Schlieren texture. The loss of the homeotropic region is a tell-tale sign that the directors must have tilted and so are no longer perpendicular or parallel to the polarizers' axes.

The polarizing microscope, therefore, is a very powerful technique for identifying liquid crystal phases, but the unambiguous assignment of the more highly ordered smectic phases requires the characterization of well-aligned monodomain samples using X-ray diffraction.

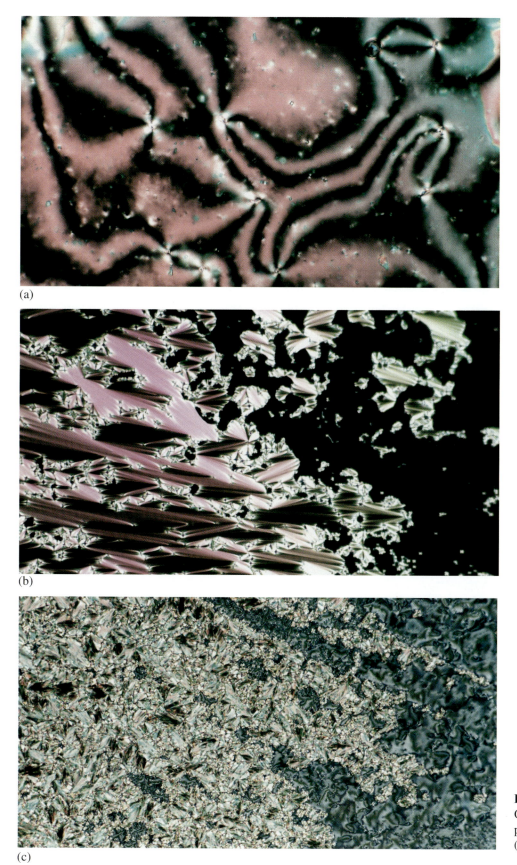

(a)

(b)

(c)

Figure 5.1
Optical textures of (a) the nematic phase, (b) the smectic A phase, and (c) the smectic C phase.

MOLECULAR STRUCTURE AND LIQUID CRYSTALLINITY

6

Now we know what liquid crystals are and how to identify them. The key question for chemists is whether we can design new liquid crystals with tailored properties in a predictable manner. To do so requires the relationship between molecular structure and liquid crystallinity first to be established and then to be understood. The pioneer in this area was a German synthetic organic chemist, Daniel Vorländer at the University of Halle. In the 1920s, he published several definitive overviews that underpinned much of the research that was to follow. Indeed, it is a great testament to these early studies that many of Vorländer's conclusions are still valid today.

One of his main conclusions was that liquid crystals are composed of rod-like molecules, and that this shape is normally achieved using a rigid mesogenic core. As this core becomes longer, so the liquid crystal–isotropic transition temperature, the clearing temperature, increases. This correlation can be seen for the examples shown in Figure 6.1. If we start with a core containing two benzene rings (Figure 6.1a) separated by an ester linkage, and then insert a third benzene ring (Figure 6.1b), then the nematic–isotropic transition temperature increases by 133 °C. This dramatic increase reflects both the stronger attractive interactions between the molecules and their ability to pack together more efficiently. Replacing the third benzene ring by a more flexible ethane link (Figure 6.1c) causes the clearing temperature to fall by 195 °C, because both the attractive intermolecular forces and the molecule's ability to pack closely together have been reduced. If we add substituents onto the benzene rings, the clearing temperature falls as the size of the substituents increases, irrespective of their chemical nature, because the rod-like shape of the molecular has been reduced.

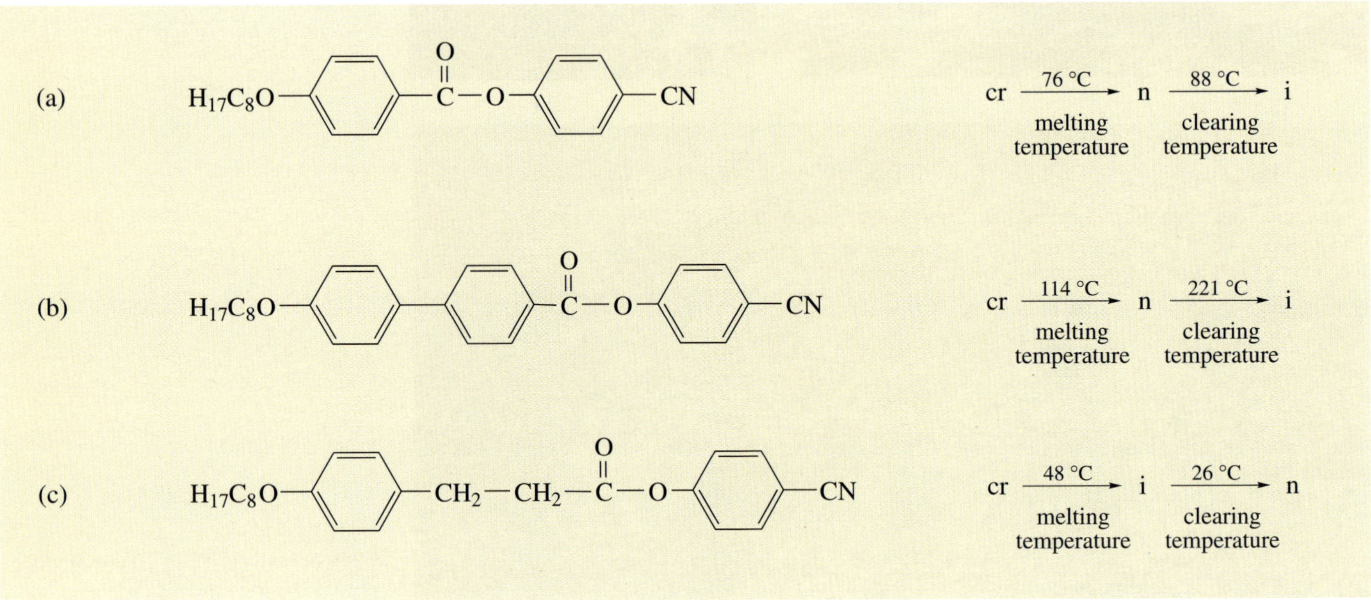

Figure 6.1

Relationship between mesogenic core structure and transition temperatures (cr, crystal; n, nematic phase; i, isotropic phase).

The discussion so far might well lead you to suppose that any molecule that deviates significantly from spherical symmetry would show liquid crystallinity. The vast majority of compounds, however, are not liquid crystals, and to understand why is very straightforward. Liquid crystals are divided into two types — *enantiotropic* and *monotropic*. An **enantiotropic** substance is one that melts into the liquid crystal phase, and subsequently clears at a higher temperature. This was exactly the behaviour observed by Reinitzer for cholesteryl benzoate. By contrast, a **monotropic** material melts directly into the isotropic liquid, and only on cooling the liquid below its melting temperature is a liquid crystal phase observable. At any moment, a monotropic liquid crystal phase can spontaneously crystallize because it exists below the melting temperature of the compound. The further you have to cool the isotropic liquid to reveal liquid crystallinity, the more likely the liquid will crystallize before the mesophase forms. Thus, if a compound's melting temperature is considerably higher than its clearing temperature, you are unlikely to see liquid crystal behaviour. In designing new materials, therefore, we must balance the intermolecular forces responsible for the three-dimensional ordering found in crystals with those responsible for the formation of liquid crystal phases. Herein lies a problem, because it is the strong interactions between the rigid cores that drives the formation of both crystal and liquid crystal phases.

We can modify the melting temperature to some extent by varying the lengths of the alkyl chains (Figure 3.1). We can think of the chains as diluting the interactions between the cores, so increasing the length of the alkyl chains tends to reduce melting temperatures. Simultaneously, we see an increased preference for smectic phases rather than nematic behaviour as the length of the alkyl chains increases. This arises because the chains and the cores are incompatible, and so do not like to mix; they can achieve this by self-assembling into layers in which the alkyl chains form one region and the cores form another.

LIQUID CRYSTAL DISPLAY DEVICES

7

The level of interest in liquid crystals declined markedly following the initial flurry of activity early in the twentieth century. Interest had waned to such an extent that by the 1960s one standard university chemistry textbook dismissed liquid crystals as being 'uncommon and of no practical importance'. Around that time, however, industrial interest began to focus on the potential applications of liquid crystals in display devices. This was the beginning of what has become — in a remarkably short period of time — a multi-billion dollar industry.

We have to return to the 1920s and 1930s to trace the origins of liquid crystal displays. Researchers in a number of countries were looking at the effects of electric and magnetic fields on liquid crystals. It had been known for some time that liquid crystals tended to align parallel to glass surfaces because of surface forces, which at the time were thought to be electrostatic in nature. Interest now lay in discovering whether an external magnetic field could achieve the same aligning effect. A very significant breakthrough occurred in 1926 when the Russian scientist Vsevold Fréedericksz showed that he could use a magnetic field to make a sample of liquid crystal sandwiched between glass plates, with its molecules aligned parallel to the surfaces, switch to a perpendicular alignment. This switch in alignment of the liquid crystal from parallel to perpendicular is known as the *Fréedericksz transition*; the analogous effect, in which the driving force is an electric field, is central to the operation of liquid crystal displays.

We now jump forward in time to the 1960s, a period of rapid technological advancement, in which electronic devices were ever shrinking, both in terms of size and cost. In particular, devices for processing information were becoming more and more widespread. But there was a problem: how was the information to be displayed? The only device at that time for displaying images was the cathode ray tube, but this was large and cumbersome, and required high voltages. By contrast, the electronic devices were small, and could work on low voltages. Clearly, there was an incompatibility between these technologies; the cathode ray tube was likened to 'a stagecoach in an era of fast cars'. What was needed was a flat display capable of operating at low voltages. From a field of contenders, liquid crystals eventually emerged triumphant.

The first generation of liquid crystal displays, developed towards the end of the 1960s, proved to be unreliable, largely due to the chemical decomposition of the liquid crystal chosen. But in 1970 the second generation of liquid crystal displays was devised by two groups working independently — in Europe, by Martin Schadt and Wolfgang Helfrich at F. Hoffmann La Roche in Basel, and in America, by Jim Ferguson working at the International Liquid Crystal Company in Ohio. The second-generation displays were called *twisted nematic liquid crystal displays* (TN-LCD; see Box 7.1); the overwhelming majority of liquid crystal displays to this day are still based on this technology.

Box 7.1 Operation of the twisted nematic liquid crystal display

In a TN-LCD (Figure 7.1) a thin layer of a nematic liquid crystal is sandwiched between sheets of glass coated with indium tin oxide (ITO). The ITO layers are transparent, and serve as electrodes, so permitting an electric field to be applied across the cell. The electrodes in turn are coated in a polyimide (the imide it is based on is shown as Structure **7.1**), whose function is to align the liquid crystal in a predetermined manner. This can be achieved by rubbing the polymer in a given direction, which causes the liquid crystals to align along that direction and parallel to the surface. Why this happens remains one of the great mysteries of liquid crystal science! The cell is constructed so that the two aligning layers have been rubbed in directions at right-angles to each other. This causes the directors in the nematic phase to twist through 90° on passing through the cell. Light entering the cell passes through a polarizer, and light leaving the cell passes through a second polarizer rotated by 90° with respect to the first (Figure 7.1a). Thus, plane-polarized light enters the cell,

R^1

R^2 R^3 =N

7.1

is guided through 90° by the twisted nematic phase, after which it can then pass through the second polarizer. In this arrangement the cell appears bright: this is the 'off' state. If we now apply an electric field (Figure 7.1b), the aligning surface forces are overcome, and the directors become parallel to the field direction. The liquid crystal layer can no longer twist the incoming light through 90°, and it is therefore blocked by the second polarizer. The display now appears black: this is the 'on' state. On switching off the electric field, the surface forces restore the twisted nematic arrangement, and the cell becomes bright again. Among the many advantages of TN-LCDs over competing technologies are lower power consumption, compatibility with low-voltage and low-power integrated circuits, flat design, reliability, fast response times, and the ability to be used in high-information content displays such as those found in laptop computers (see Figure 7.2, for another example). A notable improvement on the original design has been the use of larger twist angles, giving what are called the 'supertwist' displays.

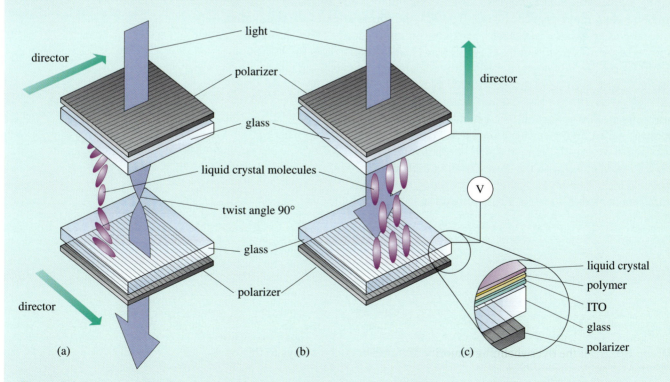

Figure 7.1 Schematic diagram, showing the design and operation of the twisted nematic liquid crystal display device (TN-LCD): (a) off-state; (b) on-state; (c) enlargement of the plate composition.

A very significant problem existed in 1970, however, which prevented the commercialization of the TN-LCD: there simply wasn't a liquid crystal with the correct combination of properties to be used in the device; in other words, the vision of the technologists could not be realized because the chemists had not provided them with the suitable materials. First of all, the liquid crystal had to align parallel to an applied electric field. In essence, this means that the molecule needs to possess a strong dipole along its principal symmetry axis*. In addition, the material had to show a nematic phase over a working range of −10 °C to +60 °C, have very high chemical and ultraviolet light-stability, be electrochemically stable, and have a low viscosity. This was perhaps the most demanding shopping list ever given to a synthetic chemist!

The daunting challenge was taken up by a Scottish chemist, George Gray, at the University of Hull, who was awarded a grant by the Ministry of Defence to work on 'Substances exhibiting liquid crystal states at room temperature'; the value of the grant was not to exceed £2117 per annum. It was this research that led directly to today's multi-billion dollar display industry! It is hard to believe that a better investment could, or ever will, be made. Gray had been active in liquid crystal science since the early 1950s, and his work was devoted to understanding the relationship between molecular structure and liquid crystallinity.

The race had begun in 1970 to create these new liquid crystals, and it was late in 1972 that the breakthrough came. Gray and his colleagues designed and synthesised the 4-alkyl- and 4-alkoxy-4'-cyanobiphenyls (Figure 7.3a and b). Members of these series of compounds proved to have the required properties, including being nematic at room temperature; however, no single compound provided the sought-after operating temperature range. Mixtures of the compounds were studied, but still the desired operating range proved elusive. By 1973, the formulation of these mixtures was proving to be a rather haphazard, time-consuming operation. This changed abruptly when Peter Raynes at the Royal Radar Establishment (now the DERA) in Malvern, Worcester, developed a method for calculating the transition temperatures of mixtures, which revealed that no mixture of the biphenyls would operate over the required temperature range. Gray argued that the temperature range could be extended by adding a component with a high clearing temperature. Gray's team synthesised 4-pentyl-4''-cyano-1,1':4,1''-terphenyl (Figure 7.3c), which they used along with the corresponding biphenyls to formulate a mixture with an acceptable operating range. This mixture was named E7, and its availability allowed the TN-LCD to be commercialized.

Figure 7.2
Mobile phone with liquid crystal display (LCD).

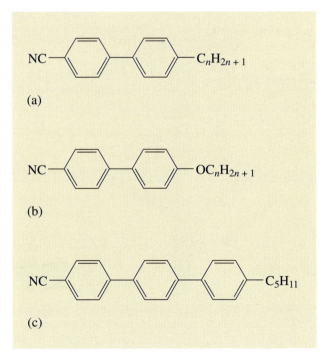

Figure 7.3 Structures of the first commercially important liquid crystals: (a) 4-alkyl-4'-cyanobiphenyls, (b) 4-alkoxy-4'-cyanobiphenyls; (c) 4-pentyl-4''-cyano-1,1':4,1''-terphenyl.

* Symmetry terms such as this are discussed in *Molecular Modelling and Bonding* 1.

CHIRALITY AND LIQUID CRYSTALS

8

How much of Reinitzer's original observations do we now understand? The opaque liquid and double melting temperature have been explained, but what of the colour changes observed on cooling? We have overlooked so far that cholesteryl benzoate is a chiral compound (Structure **8.1**). However, this manifests itself in the structure of the liquid crystal phase. Thus, cholesteryl benzoate liquid crystals have a structure we have not yet discussed, and which is known as the **chiral nematic** or **cholesteric phase**. The local molecular organization in the chiral nematic phase is indistinguishable from that found in the nematic phase, but the directors trace out a helix as you pass through the phase (Figure 8.1a). The pitch length (Figure 8.1b) of the helix is typically in the range 300–800 nm. The helical structure within the chiral nematic phase is formed because the intermolecular forces between chiral molecules are themselves asymmetric. Thus, parallel arrangements of the molecules are no longer energetically most favourable; instead, the molecular long axes are arranged with a very slight tilt angle. This almost imperceptible twisting at the molecular level reveals itself over much larger distances as a twist of the director.

When the pitch length of a chiral nematic phase is comparable to the wavelength of visible light, only particular wavelengths (or colours) are reflected, in accordance with the Bragg diffraction equation (p. 36). The interactions between the molecules are temperature

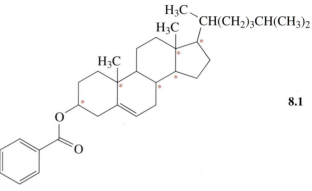

8.1

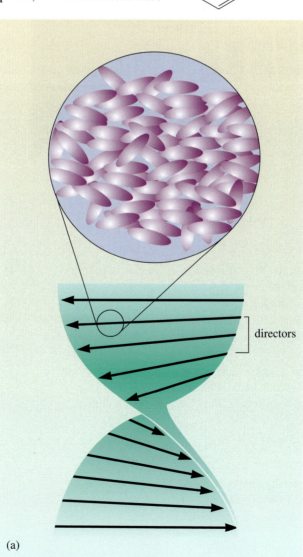

directors

(a)

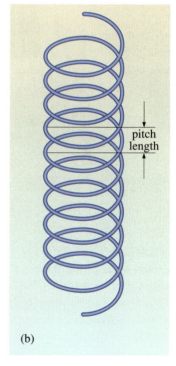

pitch length

(b)

Figure 8.1
(a) Sketch of the molecular organization within the chiral nematic phase; the arrows represent the twisting of the directors, and the inset reveals that the local organization is essentially identical to that found in the nematic phase.
(b) The pitch length of a helix is the distance along the principal axis between two identical points on the helix.

dependent, and hence the pitch length is also sensitive to temperature changes. In fact, the pitch length increases on cooling, so longer wavelengths of light are reflected as temperature falls; the accompanying colour change is from blue (hot) to red (cold). This colour change is exploited in temperature-sensing applications, ranging from children's thermometers (Figure 8.2) to the use of liquid crystals as coatings on aircraft to monitor hot spots. In a liquid crystal thermometer, the chiral nematic liquid crystal is encapsulated in a polymer matrix, and sandwiched between a transparent surface and a black support backing. This absorbs any light that might otherwise reappear through normal reflection.

Figure 8.2 Chiral nematic LCD thermometer.

The final observation of Reinitzer's still to be explained is the deep violet–blue colour that swept across the clear liquid. The chiral nematic phase adopts a single twist structure (Figure 8.1a), but in fact there is a structure that — at least at a local level — is energetically more favourable. This is a very complex structure known as a 'double twist cylinder'; these cylinders can pack together in a cubic arrangement. The spacings of the defects of this phase can be similar to the wavelength of light and they can therefore diffract light, so appearing coloured. The first such phase to be identified diffracted blue light, and hence became known as the 'blue phase'. These phases exist for very short temperature ranges, often less than a degree, before forming a chiral nematic phase.

DISCOTIC LIQUID CRYSTALS

We have seen that molecular shape is critical in forming thermotropic liquid crystal phases, and that the majority of liquid crystals are rod-like. An alternative approach was suggested in the 1920s by Vorländer, who suggested that 'flake-like' molecules ought to be able to pack face-to-face into piles. Vorländer searched for compounds that exhibited such behaviour, but he failed to find any. Indeed, it was not until 1977 that this proposition was verified by the discovery made by Sunil Chandrasekhar and his colleagues at the Raman Research Institute in Bangalore. They showed that the disc-like hexaalkanoyloxybenzenes (Figure 9.1a) are liquid crystals. As a result of their shape, these became known as **discotic liquid crystals**, and since the first discovery in 1977, a great many others have been identified.

The simplest phase seen for these disc-like molecules is the discotic nematic phase (Figure 9.1b), and this is very easy to visualize. Take a handful of coins and throw them into a box (Figure 9.1c). The coins all lie flat, and are arranged randomly, but the main axes of the coins are all arranged in the same direction; this is the discotic nematic phase. Instead of forming smectic phases, disc-like molecules exhibit columnar phases, in which the molecules are arranged into columns, which differ according to the packing arrangements within and between the columns (Figure 9.1d).

Unlike calamitic liquid crystals, discotic materials still have to find commercial application, although one promising area involves using them as environmental gas sensors. The molecules in the columns are packed so closely that their electron clouds overlap; the result is that the electrons are able to move up and down the stack. If a gas molecule is absorbed into the column, the spacing between the discs changes, and, as a consequence, so will the conductivity of the column. This process provides a means of detecting gases.

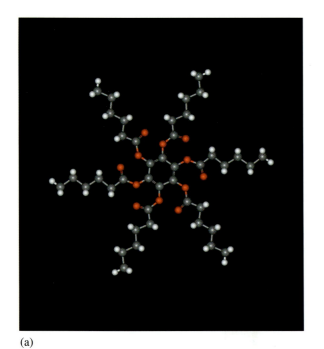

(a)

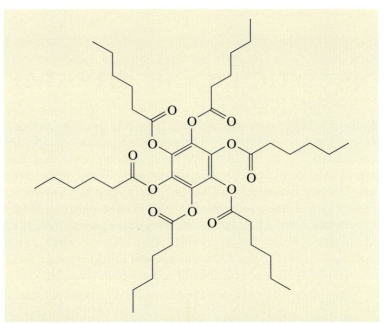

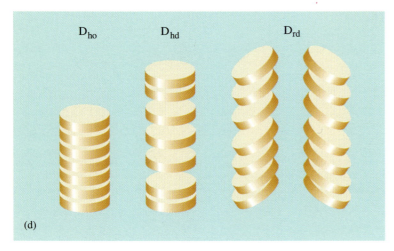

(b)

(c)

D_{ho} D_{hd} D_{rd}

(d)

Figure 9.1

(a) WebLab ViewerLite and line drawing of the structure of the molecules in the first discotic liquid crystals: the hexaalkanoyloxybenzenes.

(b) Schematic diagram of the molecular organization found in the discotic nematic phase.

(c) Layers of coins.

(d) Columnar liquid crystal phases: the arrangements in the D_{ho}, D_{hd} and D_{rd} forms are illustrated. 💻

LIQUID CRYSTAL POLYMERS

10

Again we return to Vorländer, who in the 1920s posed the question 'what would happen to the molecules (liquid crystals) when they become longer and longer?'. So was born the field of liquid crystal polymers. He showed that short chains — or oligomers — of 4-hydroxybenzoic acid were liquid crystals, but as the chain length increased, the polymer quickly became intractable, decomposing before melting. During the 1930s, solutions of the rod-like tobacco mosaic virus particles were found to be birefringent above a critical concentration, and 'softening points' were reported for alkyl-celluloses. With hindsight, these are the earliest observations of liquid crystal polymers, but they were not identified as such at the time.

Figure 10.1
Logs floating down a river.

In the 1950s, Lars Onsager and Paul Flory predicted that solutions of rod-like molecules would spontaneously order above a critical concentration, which depended on the length-to-breadth ratio of the rod (or macromolecule). This is easy to imagine by considering logs floating down a river (Figure 10.1). If there are relatively few logs, they arrange themselves in a rather haphazard fashion, but as the number of logs is increased they all tend to align themselves in the same direction, forming liquid crystalline-like order. This gives us a model of a **lyotropic liquid crystal** (from the Greek *lyo* 'to dissolve'). At the same time, scientists at Courtaulds in Maidenhead found that concentrated solutions of poly(methyl glutamate) (Figure 10.2a) and poly(benzyl glutamate) (Figure 10.2b) in, for example, dioxane and dichloromethane solution, respectively, were liquid crystalline. At first sight, these may not appear to be rod-like macromolecules, but in solution the molecules adopt extended helical secondary structures, and so behave like long, rigid rods. These rods pack together efficiently, with their long axes essentially parallel.

In the early 1970s, Stephanie Kwolek and her colleagues at DuPont found they could make fibres of exceptionally high tensile strength by processing solutions of rod-like polymers, identified only later as liquid crystals. The first product they

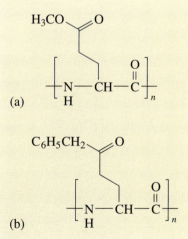

Figure 10.2
Molecular structures of:
(a) poly(methyl glutamate);
(b) poly(benzyl glutamate).

developed, called Fibre B, was based on poly(benzamide), but this was soon superseded by poly(*p*-phenylene terephthalamide) (Figure 10.3), known commercially as kevlar. This material was developed independently by the Dutch company Akzo Nobel, who marketed it as Twaron.

Figure 10.3
Molecular structure of poly(*p*-phenylene terephthalamide), known commercially as kevlar or Twaron.

The exceptionally strong mechanical properties of kevlar fibres can be attributed to the way in which the liquid crystal solutions are processed (Figure 10.4). A liquid crystalline solution of kevlar in fuming sulfuric acid (concentrated sulfuric acid containing dissolved sulfur trioxide — the only solvent capable of dissolving kevlar!), consisting of randomly arranged liquid crystal domains, is pushed through a set of small holes called a spinneret. As the solution passes through the holes, shear forces align the liquid crystal domains in the flow direction. On emerging from the spinneret, however, the fibre swells, reducing the degree of ordering of the macromolecules. In order to recover this ordering — and indeed to enhance it — the fibre is stretched. We now have a fibre in which the macromolecules are aligned along the fibre axis; this arrangement is fixed by coagulating the fibre in water. On a weight-for-weight basis, kevlar fibres processed in this way are ten times stronger than steel, and because of this they find application in diverse areas including anti-

Figure 10.4
Processing solutions of kevlar.

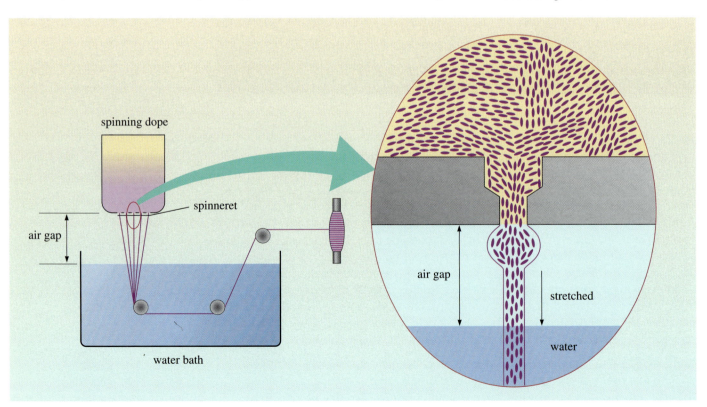

ballistic garments (bullet-proof vests), high-strength textiles for sails, rubber reinforcement in radial tyres, body panels for cars, and high-strength ropes (Figure 10.5). Kevlar's resistance to corrosion by seawater makes it suitable for use in the offshore oil and gas industry.

Kevlar is an example of an aramid (aromatic amide), and such compounds still dominate the industrially important liquid crystal polymer sector. These materials are not without problems, however, not

Figure 10.5
Kevlar is used for high-strength applications, as in this glove.

least that aggressive solvents such as fuming sulfuric acid have to be used to process them. This provided the driving force to find thermotropic liquid crystal polymers — polymers that melt into a liquid crystal phase, and which can therefore be processed from the melt.

The principal explanation for the aramids' extreme unreactivity and insolubility is the strong hydrogen-bonding between the polymer chains. Only solvents that disrupt these hydrogen bonds can dissolve the polymer. The amide linkages therefore have to be replaced by a unit not capable of undergoing hydrogen-bonding; an ester linkage is most often used. In principle, the intermolecular forces are considerably weaker, and the polymer should melt prior to decomposing. The aromatic polyester analogue of kevlar, poly(p-phenylene terephthalate), decomposes before its melting temperature is reached. Thus, not only must the strength of the intermolecular forces be reduced, but also the packing efficiency of the macromolecules in the solid state. Consequently, two quite different design strategies emerged, giving either rigid or semi-flexible main-chain liquid crystal polymers.

Industrial interest has tended to focus more on the rigid main-chain polymers because highly orientated samples posess excellent mechanical properties. The most successful commercial polymer, called Vectra, was developed by Hoechst Celanese (Figure 10.6). In this material the packing efficiency of the chains is reduced by the incorporation of a 'crankshaft' monomer, 6-hydroxynaphthalene-2-carboxylic acid (Structure **10.1**). The polymer has a low melt viscosity, and there is a strong tendency for the macromolecules to align in the flow direction. These materials can therefore be used for injection moulding of parts with complex shapes. Apart from their excellent strength, the polymers show good chemical and thermal stability, and have very low thermal expansion properties. These properties are exploited in a number of applications, including mounts for the optical components of compact disc players, fibre optic connectors, microwave cooking utensils, and in filament-wound pressure vessels for fire extinguishers. NASA is also evaluating their use as part of a thermosetting resin to be used as a protective coating on the fuel tanks of the second-generation space shuttle.

HOOC ⎯ OH

10.1

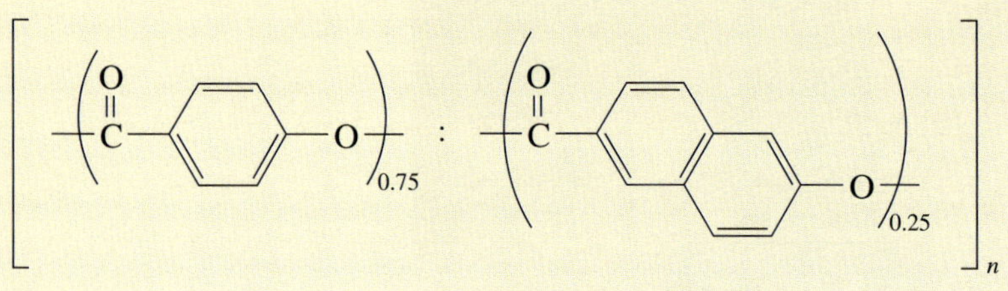

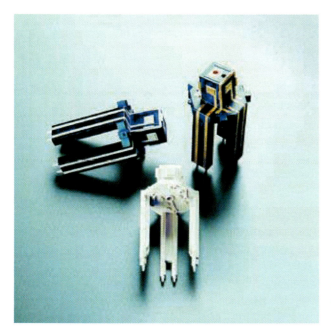

Figure 10.6   The structure of the thermotropic liquid crystal polymer, Vectra.

Figure 10.7
Sunlight sensors made from Vectra.

However, the widespread use of these aromatic polyesters is somewhat restricted by their high price. Attempts to recycle used Vectra have met with only partial success because of the concomitant degradation of its mechanical properties. As a result, their use in large-scale structural applications is always going to be rather unlikely.

SIDE-GROUP LIQUID CRYSTAL POLYMERS

11

The other main class of liquid crystal polymers is the side-group liquid crystal polymers, which were discovered independently in the late 1970s by two research groups — a team at the Moscow State University led by Valery Shibaev, and a team at the University of Mainz led by Helmut Ringsdorf. In these polymers, the mesogenic liquid crystal groups, either rods or discs, are attached as pendants onto a conventional polymer via a flexible spacer (Figure 11.1a). The insertion of the flexible spacer between the mesogenic group and the backbone (tinted part of Figure 11.1b) is critical; without it, the material is not normally liquid crystalline. The role of the spacer is to decouple the opposing tendency of the mesogenic groups to self-assemble into a liquid crystal phase from that of the backbone to adopt random coil arrangements. This decoupling endows a unique duality of properties on these polymers; specifically, they show macromolecular characteristics such as glass-like behaviour and mechanical integrity, combined with the electro-optic properties of low molar mass liquid crystals, albeit on a much slower time-scale. It is this duality of properties that underpins much of the exciting application potential of side-group liquid crystal polymers.

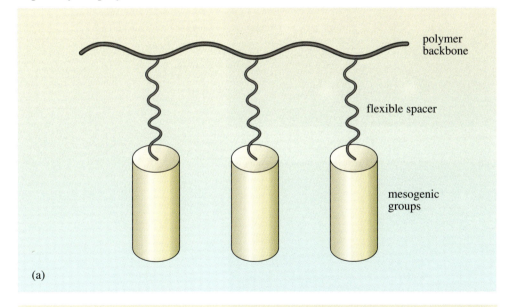

Figure 11.1
(a) Schematic representation of a side-group liquid crystal polymer; (b) a typical example based on polystyrene and containing a rod-like liquid crystal group.

One of the most extensively investigated applications of side-group liquid crystal polymers is optical information storage based on photoisomerization. This requires a polymer containing groups that change their shape when irradiated with light of a particular wavelength; the most widely used photoisomerizable unit is the azobenzene group. In the *trans*-form (Figure 11.2a), azo groups enhance the rod-like shape of the liquid crystal unit, and hence promote liquid crystallinity. On irradiation with light of a given wavelength, the azo group isomerizes to the bent *cis* isomer (Figure 11.2b), which militates against the organization necessary for a liquid crystal phase. Several devices have been built around this general effect. In one, developed by Tomiki Ikeda at the Tokyo Institute of Technology, a strongly light-scattering opaque film is obtained by cooling the polymer from a liquid crystal phase (Figure 11.3). On irradiation with a laser, the *trans–cis*-isomerization results in the formation of areas of isotropic phase which are optically transparent. It is unclear exactly how the liquid crystal–isotropic transition is achieved, but it is

Figure 11.2
The change in configuration of azobenzene from (a) the linear *trans*-form to (b) the bent *cis*-isomer on irradiation. ▱

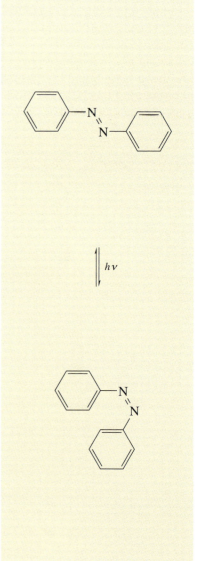

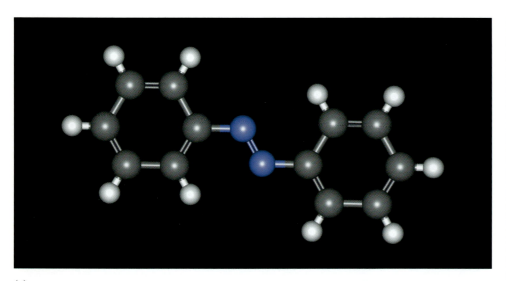

(a)

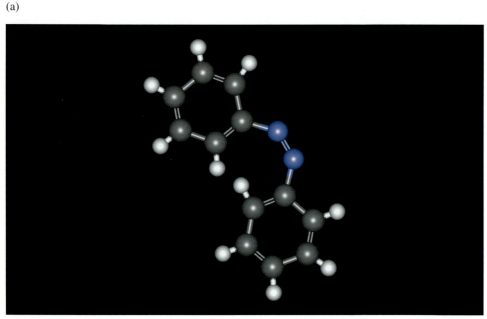

(b)

generally considered to occur without change of temperature, and can happen even below the glass transition temperature of the polymer. The thermal back transition — that is, *cis*- to *trans*- — is relatively rapid, but providing that the polymer is below its glass transition temperature, the side-groups are unable to reorganize to form the liquid crystalline phase. Thus, these small transparent regions are essentially stored indefinitely on an opaque background. To erase this information, the area simply requires heating above the glass transition temperature; the reorganization of the side-groups to form an opaque liquid crystal phase occurs very rapidly.

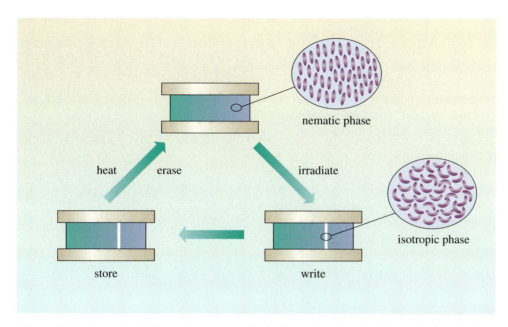

Figure 11.3 An optical information storage device based on the photoisomerization of, for example, an azobenzene-based liquid crystal.

12

LYOTROPIC LIQUID CRYSTALS

We now turn our attention to the second major class of liquid crystals, **lyotropic liquid crystals**. These are composed of **surfactants** or **amphiphiles** (from the Greek *amphi* 'both' and *phil* 'love', so loving both kinds), which are molecules containing two distinct regions, one polar and the other non-polar; the polar region, or head-group, might be a **hydrophilic** ionic group such as phosphate, and the non-polar region, or tail, a **hydrophobic** alkyl chain. Liquid crystallinity is observed when these amphiphiles are dissolved in an appropriate solvent. The most common examples of lyotropic liquid crystals are soaps and detergents dissolved in water; in fact, the slime you find in your soap dish is a liquid crystal phase. Every time you wash your hands or do the dishes, liquid crystalline phases form in the dirty water. Indeed, it could be argued that the first liquid crystals to find application were the soaps used thousands of years ago by, among others, the Phoenicians.

A typical phase sequence for a lyotropic liquid crystal in water is shown in Figure 12.1. At very low concentrations, individual molecules are surrounded by solvent molecules (Figure 12.1a). Then, at some critical concentration, known as the *critical micelle concentration*, the amphiphilic molecules spontaneously form aggregates, or **micelles**, in which the polar parts of the molecules are on the outside of the micelle and are solvated, whereas the non-polar regions are inside and shielded from the solvent (Figure 12.1b). The formation of these spherical micelles is driven by the hydrophobic effect, which acts to squeeze the oil or hydrophobic regions out of the water. On increasing the surfactant concentration, the micelles become increasingly densely packed, and interact via repulsive electrostatic forces. At still higher concentrations, the micelles must change their shape in order to accommodate the greater number of molecules; rod-like micelles are then formed. These close-pack in a hexagonal arrangement to give a phase in which the rods are separated by water (Figure 12.1c). As the water content is reduced further, the rods transform into bilayers, to give a phase in which they are stacked with water in between (Figure 12.1d). This is the lyotropic analogue of the smectic A phase. If the water content is reduced still further, the inverse forms of these phases can form, in which regions of water are now enclosed and the hydrophobic groups are on the outside of the aggregates. It is of considerable importance to industry to study these changes, as use is made of the favourable characteristics of liquid crystalline phases in processing soaps and detergents.

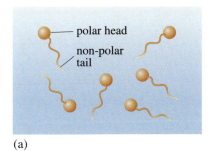

(a)

Figure 12.1 Schematic representations of lyotropic liquid crystal phases. On increasing the concentration of the amphiphile, the structure changes from (a) isolated molecules to (b) spherical micelles, to (c) rod-like micelles packed in a hexagonal arrangement, and eventually to (d) a lamellar phase.

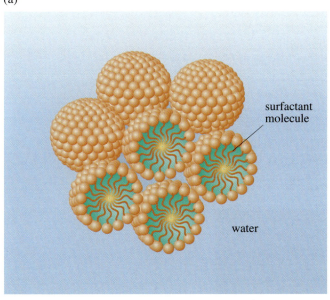

(b)

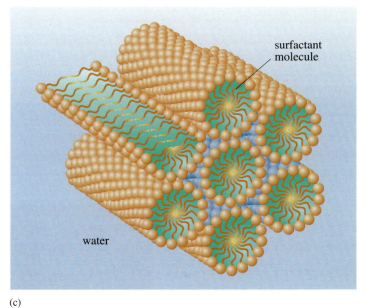

(c)

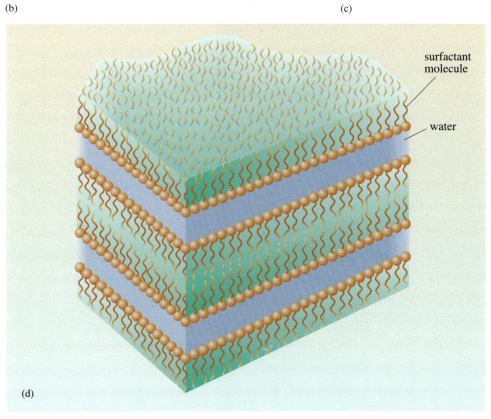

(d)

LIQUID CRYSTALS AND LIFE

13

Some 30 years before Reinitzer's observation of the extraordinary melting behaviour of cholesteryl benzoate, a German ophthalmologist called Mettenheimer reported that the nerve sheath protein myelin was birefringent, and that it had fluid properties; many years later, this was ascribed to liquid crystalline behaviour. Myelin contains regularly stacked layers of membranes, which are found everywhere in living systems; for example, cell walls are composed of amphiphiles such as dioleoylphosphatidylcholine (Figure 13.1). In the presence of water, the amphiphiles form bilayers (like the lamellar phase shown in Figure 12.1d), which stack on top of each other, giving a myelin-like structure. The molecules are free to move around the membrane, but they cannot move between membranes as this would expose hydrophobic regions of the molecule to water. Such membranes are therefore often described as two-dimensional fluid fabrics. As we shall see, this fluidity is critically important in that it allows proteins embedded in the membrane to move around.

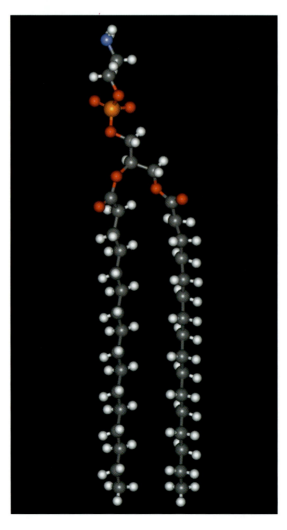

Figure 13.1
Line drawing and WebLab ViewerLite image of dioleoylphosphatidylcholine, a naturally occurring glycerophospholipid and a component of membranes in living systems. 🖳

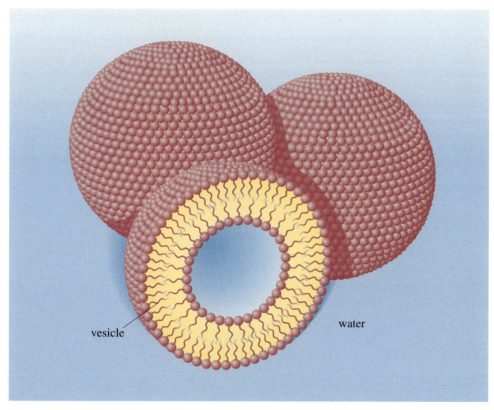

Figure 13.2 A sketch of a vesicle.

If more water is added to the system, the bilayers break down and globular sacs called **vesicles** are formed, with a wall comprising a single bilayer (Figure 13.2). Each vesicle contains a small volume of water, isolated from the surrounding water. These structures are thought to have played a key role in the evolution of life on Earth, representing primitive cells. They are now known to be involved in the transfer of proteins between organelles. It is the liquid-crystalline nature of the membrane which gives the cell the flexibility to divide, and it provides the proteins embedded in the membrane with mobility, thus ensuring that new cells have the same proteins.

Other liquid crystal phases are also thought to play an important role in biological processes. Phospholipids extracted from the brain show hexagonal phase stacking (cf. Figure 12.1c), although the significance of this is not yet fully understood. More complicated structures, the cubic phases, have been found in nature during the process of cell division. Naturally occurring biopolymers such as DNA also show liquid crystallinity. Again, the importance of this is as yet unclear.

WHAT NEXT?

14

Liquid crystal displays will continue to have a pre-eminent role in the display industry, especially in association with wireless technologies such as laptop computers and mobile phones. New display technologies based on chiral liquid crystals are likely to challenge the cathode ray tube, since flexible displays that can be rolled up like a piece of paper have been shown to be possible. It is also apparent that the applications of liquid crystals will become even more diverse, encompassing such areas as information storage, molecular wires and sensors. Underpinning all of this applied research, fundamental work will have to be carried out on how and why molecules self-assemble. This, in turn, will lead to a greater understanding of the role of liquid crystals in nature, and to the design of new materials for biological and pharmaceutical applications. Liquid crystal science will undoubtedly be a tremendously exciting area to be involved in for a long time to come.

FURTHER READING

1   E. A. Moore, *Molecular Modelling and Bonding*, The Open University and the Royal Society of Chemistry (2002).

ACKNOWLEDGEMENTS

Figures

Figure 1.1b: *Age of the Molecule,* edited by Nina Hall, Royal Society of Chemistry (1999); courtesy of University of Aberdeen; *Figure 5.1*: Corrie Imrie; *Figure 8.2*: courtesy of Sharn Inc; *Figure 10.1*: Heather Angel; *Figure 10.5*: E. I. Dupont de Nemours and Company; *Figure 10.7*: Hoechst.

INDEX

Note Principal references are given in bold type; picture references are shown in italics.

CD-ROM INFORMATION

Computer specification

The CD-ROMs are designed for use on a PC running Windows 95, 98, ME or 2000. We recommend the following as the minimum hardware specification:

processor	Pentium 400 MHz or compatible
memory (RAM)	32 MB
hard disk free space	100 MB
video resolution	800 × 600 pixels at High Colour (16 bit)
CD-ROM speed	8 × CD-ROM
sound card and speakers	Windows compatible

Computers with higher specification components will provide a smoother presentation of the multimedia materials.

Installing the CD-ROMs

Software must be installed onto your computer before you can access the applications. Please run INSTALL.EXE from either of the CD-ROMs.

This program may direct you to install other, third party, software applications. You will find the installation programs for these applications in the INSTALL folder on the CD-ROMs. To access all the software on these CD-ROMs, you must install QuickTime, Isis/Draw, WebLab ViewerLite and Acrobat Reader.

Running the applications on the CD-ROM

You can access *The Third Dimension* CD-ROM applications through a CD-ROM Guide (Figure C.1), which is created as part of the installation process. You may open this from the **Start** menu, by selecting **Programs** followed by **The Molecular World**. The CD-ROM Guide has the same title as this book.

The *Data Book* is accessed directly from the **Start | Programs | The Molecular World** menu (Figure C.2).

Problem solving

The contents of this CD-ROM have been through many quality control checks at the Open University, and we do not anticipate that you will encounter difficulties in installing and running the software. However, a website will be maintained at

http://the-molecular-world.open.ac.uk

which records solutions to any faults that are reported to us.

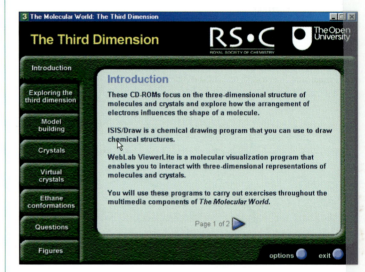

Figure C.1 The CD-ROM Guide.

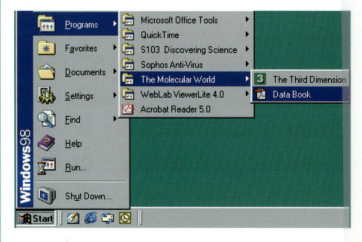

Figure C.2 Accessing the *Data Book* and CD-ROM Guide.